GUIDE TO THE PLANETS

GUIDE TO THE PLANETS

by

Patrick Moore, O.B.E., F.R.A.S.

LUTTERWORTH PRESS · LONDON

This completely revised and new edition first published 1971

ISBN o 7188 1731 1

Printed in Great Britain by
Ebenezer Baylis and Son Ltd.
The Trinity Press, Worcester, and London

CONTENTS

LIST OF PLATES

(Plates I to XVI between pages 112 and 113)

I. VENUS
   (left) Photograph with Palomar 200-in. reflector
   (right) Photograph in blue light with Mount Wilson
   100-in. reflector

II. THE MOON
   Photographic section showing Mare Crisium and Mare
   Tranquillitatis

III. LUNAR SCENERY
   (above) The area of Aristarchus
   (below) The Sinus Iridum

IV. LUNAR SCENERY
   (left) The Straight Wall and the craters Ptolemæus,
   Alphonsus and Arzachel
   (right) A crowded upland area in the south part of the
   Moon including the crater Tycho

V. LUNAR SCENERY
   (above left) Gassendi and Mare Humorum
   (above right) Plato and the Alpine Valley
   (below left) Part of the Mare Nectaris, with the Altai
   Scarp
   (below right) Grimaldi and Riccioli area on the Moon's
   limb

VI. THE EARTH FROM APOLLO 8

VII. MARS
   (above) Mars in blue and red light
   (below) Mars and its satellites

7

CONTENTS

FOREWORD

THE ORIGINAL EDITION of *Guide to the Planets* was published twenty years ago, at a time when the idea of travelling to the Moon was officially regarded as ridiculous. Several subsequent editions appeared over the next decade, but the last of these, with the shortened title of *The Planets*, came out in 1962, and much has happened since then. For the present edition, I have re-written the entire text; the illustrations also are mainly new, and I have done my best to include the latest information, even though in these stirring times one is always apt to be overtaken by the march of events!

PATRICK MOORE

Selsey, *July 1971*

9

ACKNOWLEDGEMENTS

MANY PEOPLE have helped in the compilation of this book. The line diagrams were drawn by Miss Patricia Cullen, to whom I am extremely grateful. Photographs were made available to me by Commander Henry Hatfield, W. Rippengale and R. Smith; and as always, I have received the utmost help and courtesy from Michael Foxell and all his colleagues at Lutterworth Press.

Selsey.   P.M.

Chapter One

THE 'WANDERING STARS'

WE LIVE IN EXCITING times. The Earth is our home, but already some men have travelled far beyond it. A new age began on July 21, 1969, when Neil Armstrong and Edwin Aldrin stepped out on to the surface of the Moon. True, they did not stay for long, and their journey was in the nature of a pioneer reconnaissance, but it was immensely significant. There must surely be lunar bases well before the turn of the century, and I have little doubt that some of the readers of this book will themselves go to the Moon.

Of course, this is only a beginning. The Moon, at a distance of only about a quarter of a million miles, is very much our nearest natural neighbour in space, and it stays together with us as we travel round the Sun. It is not a friendly world, and nothing can ever make it so. Yet it is more than a terrestrial outpost; it represents a stepping-stone to other worlds.

Nobody should need reminding that our Earth is a planet, moving round the Sun. It is a junior but important member of the Solar System, and as planets go it is neither exceptionally large nor exceptionally small, but it is unique inasmuch as it has the sort of atmosphere which we can breathe. If this were not so, we would not be here. We would have to be in some other system, since of all the planets in the Sun's family only Mars, apart from Earth, holds out the slightest hope of being able to support life—and even this now seems depressingly doubtful.

In the present book, my aim is to give an account of the planets, those 'other Earths' which orbit the Sun at different distances and at different speeds. They make up a family which is as fascinating as it is varied. Mercury and Venus are the innermost members; then comes the Earth, and then Mars. After a wide gap, we meet with the four giants: Jupiter, Saturn, Uranus and Neptune. On the fringe of the system comes Pluto, which was discovered only forty years ago, and which remains very much of a problem world. Beyond Pluto there is nothing but thinly-spread material—until we come to the closest of the night-time stars.

Our Sun is a star. It seems brilliant and glorious to us only because it is a mere 93,000,000 miles* away, which to an astronomer is not very much. The other stars have distances which have to be measured in millions of millions of miles. Represent the Earth–Sun distance by one inch, and the nearest star, a dim red dwarf called Proxima, must be taken over four

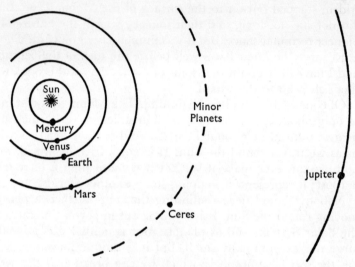

Fig. 1. Plan of the Solar System
a. The inner planets, with the orbit of Jupiter.

miles away. The Solar System is decidedly isolated in the universe.

Though the planets seem starlike when viewed with the naked eye, they are not in the least starlike in nature. A star shines by its own energy, and is losing mass in the process; our Sun is losing mass at the rate of four million tons every second—though I hasten to add that there is plenty left! A planet has no light of its own, and depends upon reflecting the light of the

* Now that Britain is threatened with 'metrication', the modern tendency is to give all distances in the Metric System. However, as yet (1971) we have not been bludgeoned into dull conformity with Europe, and I propose to retain the English system of measurement here, for the excellent reason that everybody can understand what it means.

Sun. Also, it appears as a disk when seen through a telescope, whereas no telescope yet built or planned will show a star as anything but a point of light.

Before going any further it will, I think, be helpful to give a diagram of the Solar System, putting in the planets to their correct relative distances from the Sun. Unfortunately it is hopeless to try to fit the diagram on to a single page, because the System is so large. On average, Mercury is only 36,000,000

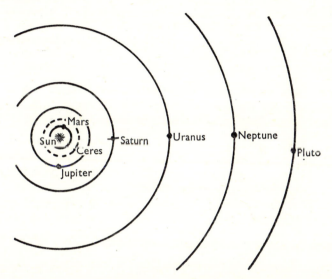

b. The outer planets, drawn to a different scale; Mars' orbit is also shown.

miles from the Sun, but Pluto is a hundred times as far, so that it is easier to give the diagram in two parts.

First (Fig. 1a) there are four relatively small planets: Mercury, Venus, the Earth and Mars, all of which are solid, and which have various points in common, so that they are often known as the terrestrial planets. Beyond Mars comes a 'natural break' in the Solar System, in which move thousands of midget worlds known variously as asteroids, planetoids or minor planets. In the second diagram (Fig. 1b), I have put in Mars and the asteroid zone to act as a link between the terrestrial planets and the four giants, which are quite unlike the Earth, and which are made up of gas (at least in their outer layers).

To make the situation clearer, I have worked out a scale model in which the Sun has been reduced to a globe 600 feet in diameter, and set down on Westminster Bridge in the heart of London. The planets can then be put in, as follows:

MERCURY: a globe 2 feet in diameter, 4½ miles away in Hampstead

VENUS:	5¼	9	Mill Hill
EARTH:	5½	12¼	Barnet
MARS:	3	18½	St. Albans
JUPITER:	60	64	Northampton
SATURN:	51	117	Lincoln
URANUS:	20	236	Sunderland
NEPTUNE:	20¼	370	Montrose
PLUTO:	2½	485	Wick

The diagram (Fig. 2) is useful in another way, because it brings out the way in which the planets are spaced. Looking at a conventional plan, it is easy to think of the giants as being 'next door' to each other, but actually they are not. Drive from London to Sunderland, and by the time you have reached Lincoln you will be only about half-way. It follows that in the Solar System, Uranus is as far from Saturn as we are.

Putting in the stars to the same scale presents another problem, be-

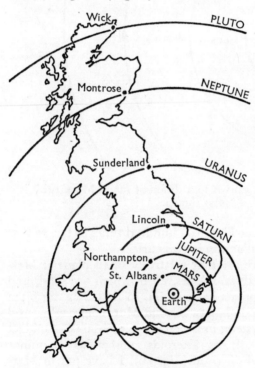

Fig. 2. Orbits of the planets, scaled with respect to Britain.

cause their distances are so vast. Reduce the Sun to a globe only one foot in diameter, and the Earth will be a bead at 36 yards; the nearest star will have to be taken and deposited somewhere in Russia. All of which shows that simply because we have sent men to the Moon, we are very far from 'conquering the universe'.

Of course, the diagrams given here are very over-simplified. For example, they show the orbits of the planets as being circular. To be accurate, they should be elliptical—though the error is not great, on this scale, except for Pluto, which can actually swing within the path of Neptune, and could come south of Montrose on our map.

The planets have different revolution periods. The Earth, as we all know, takes $365\frac{1}{4}$ days to go once round the Sun. Mercury, moving more quickly in a smaller orbit, takes only 88 days; Pluto has a 'year' 248 times as long as ours. Then, too, there are other members of the Solar System, notably the flimsy but sometimes spectacular comets, and the almost countless small particles which make up meteors and meteorites. There is also a surprising amount of what we may term 'dust', and it has been found that the space between the planets is not empty, as used to be thought not so very long ago.

Four of the planets are brilliant. Venus, Mars, Jupiter and Saturn can hardly be overlooked even by the most myopic observer. The ancient star-gazers knew them well, and realized that they are basically different from the true stars, because they wander about from one star-group to another—though they always keep to certain well-defined parts of the sky. There is no mystery about this behaviour. The true stars are moving about rapidly, but they are so remote that their individual motions are too slight to be noticed with the naked eye even over long periods. The planets are much closer to us, so that their shifts are obvious even over a night or two. The very word 'planet' really means 'wandering star'.

To these four brilliant planets we must add Mercury, which is distinctly elusive, but which was known long before the time of Christ; at times it can be quite conspicuous in the dawn or dusk sky. Of the rest, Uranus can just be seen with the naked eye, but Neptune and Pluto are too faint. Uranus was discovered in 1781, Neptune in 1846 and Pluto as recently as 1930.

Whether another planet exists, still further away, is a matter to which I shall return in Chapter 16.

The Greeks, who may be classed as the first really scientific observers, worked out the movements of the planets with considerable accuracy, even though they made the mistake of supposing that the Earth, not the Sun, lies in the centre of the system. (Occasional rebels, such as the famous philosopher Aristarchus, were officially dismissed as heretics.) It was only about three and a half centuries ago that the Earth was finally relegated to the status of an ordinary planet. When telescopes were developed, from 1609 onward, it became possible to study the features of the planets themselves. Mercury and Venus were found to show phases, similar to those of the Moon; the red surface of Mars was blotched with darker areas, and there were whitish polar caps which looked like snow; Jupiter was crossed by cloud belts, while Saturn was distinguished by its superb set of rings.

Telescopes of the kind used by modern amateur astronomers will show fine details on the planets, and until the beginning of the Space Age, in the 1950s, most of our knowledge of their surfaces was based on amateur work. We must admit that this is no longer true. Rocket probes have been sent out to Mars and Venus, and before 1980 it is likely that some of the other planets will also have been examined from close range. Yet this does not mean that observations carried out from Earth have become pointless. Each planet has its own special features, and many mysteries remain to be solved.

The main disappointment, so far, is that there seems to be a total lack of life anywhere in the Solar System except on Earth. This is unfortunate, but it does not lessen either our interest or our wish to explore; and observers have a vital rôle to play. In or around 1990, the first men will step out on to the dusty wastes of Mars. They will have immense dangers to face; the more information we can collect for them, the better will be their chances of a safe return. And although Mars must be our first planetary target, it is not too much to hope that eventually, perhaps hundreds of years in the future, our messengers will go out into the remotest depths of the Solar System.

Chapter Two

THE BIRTH OF THE PLANETS

EVERYBODY IS INTERESTED in the problem of the world's creation. Unfortunately nobody has been able to provide an answer satisfactory enough to be generally accepted, and nobody (apart from the Biblical Fundamentalists) has ever claimed otherwise. There are many ancient legends; I particularly like one told by the Iroquois Indians of North America, who said simply that 'a heavenly woman was tossed out of Heaven, and fell upon a turtle, which grew into the Earth'.

So far as the origin of the universe is concerned, we have to admit that our ignorance is complete. Starting with a universe containing material in some form or other, we can work out a sequence of events, starting with uniform gas and ending up with the Earth; but where did the material come from in the first place? We have no idea, and to delve into the question is beyond the scope of a book dealing with the planets. So let us assume that we have material to act as 'building blocks', and start from there.

The first thing to do is to decide upon some sort of time-scale, and for once we are on comparatively safe ground, because we know that the age of the Earth is about 4,700 million years. There are various lines of investigation, all of which lead to the same result. Rock samples brought back from the Moon by the Apollo astronauts and the Russian probes indicate that the age of the Moon is also about 4,700 million years, and we are fully entitled to assume that all the planets were formed at about the same time. The same applies to the minor bodies— notably the meteorites, which may land on Earth and submit themselves to analysis in our laboratories.

Recorded history does not carry us far back into the past. Archæological research can reach back further, and the study of fossils—the remains of long-dead creatures—can bring us news from hundreds of millions of years ago. Fossils, indeed, gave the first definite proof that the world is very old. Previously, the Church had been regarded as the supreme authority; in particular there was Ussher, Archbishop of Armagh, who stated in 1654 that the world had been created at nine o'clock in the morning of October 26, 4004 B.C. (He had reached this startling conclusion by adding up the ages of the patriarchs, and making other equally irrelevant calculations. I have never been able to find out whether he made due allowance for Summer Time!)

However, fossils cannot tell us about the beginning of the story, because the Earth cannot have been born suddenly. The process must have been gradual, and it seems that at first the surface was too hot to support life. A more extended method of dating is that known commonly as the radioactive clock, depending upon substances such as uranium.

Uranium, the heaviest of the ninety-two fundamental substances or elements found naturally on Earth, is not stable. It decays spontaneously, and finishes its career as lead. It is in no hurry to disintegrate; and for one type of uranium—known scientifically as U.238—well over 4,000 million years must elapse before half of the original element has become lead. (With radium, this 'half-life' is only 1,620 years.) Lead produced from uranium can be distinguished from ordinary lead, and so the quantity of uranium-lead associated with the remaining uranium tells us how long ago the decay started. This, in turn, gives a lower limit to the age of the rocks themselves.

Another radioactive element is rubidium, which ends up as strontium. Here the half-life is 46,000 million years, which is long even by cosmical standards. Some remarkable studies have been made of rubidium in the Rocks of St. Paul, a group of volcanic islands on the equator between Africa and South America. It appears that the decay has been in progress for 4,500 million years, so that the Earth must be older than this.

Not so very many decades ago, ideas of this sort would have been dismissed out of hand. Lord Kelvin, one of the great physicists of Victorian times, estimated the age of the world as little over ten million years; he worked out this figure by calculating how long the Earth would take to cool down to its present temperature if it had originally been as hot as the Sun (surface temperature 6,000 degrees Centigrade). I do not propose to discuss the matter further, because there really seems no doubt that the present estimate of 4,700 million years is very near the truth. Man is a newcomer to the terrestrial scene, and historical periods seem very short on the universal time-scale.

If we represent the age of the Earth by one day, the Battle of Hastings was fought only one second ago.

We are also fairly safe in assuming that the Earth and the other planets were formed either from the Sun itself, or from a companion star formerly linked with the Sun, or from a cloud of diffuse matter which once surrounded the Sun. All we have to do is to work out the process of formation—but it is here that our difficulties begin.

The first really scientific attempt was made in 1796 by Laplace, a famous French astronomer, who produced what is always called the Nebular Hypothesis. (It had the same basis as earlier ideas put forward by Thomas Wright in England and Immanuel Kant in Germany, but was much more detailed and plausible.) Laplace started with the picture of a vast gas-cloud, disk-shaped and in slow rotation. He then worked out an evolutionary sequence ending up in a system with a central sun attended by the planets and their satellites.

Laplace supposed that as the gas-cloud cooled down, radiating its heat away into space, it shrank; as it shrank, its rate of spin increased, until the centrifugal force at its edge became equal to the gravitational pull there. At this stage a ring of matter broke away from the main mass, and condensed gradually into a planet. As the shrinkage of the main mass continued, a second broken-off ring gave rise to a second planet; and the process was repeated over and over again, so that the end product was a central body (the Sun) surrounded by a retinue of circling worlds. The outermost planets were the oldest, and Mercury, the closest planet to the Sun, was the youngest member of the solar family. (Laplace actually began with Uranus, since in 1796 the two most distant planets, Neptune and Pluto, were not known.)

At first sight the Nebular Hypothesis looks very neat, and it was accepted for many years, but it has not stood up to the severe tests of mathematical analysis. One trouble is that the material shed during the shrinkage would not form separate rings—and even if it could, such a ring would not condense into a planet. Another objection often put forward was that since the original gas-cloud was presumably flat, the orbits of the planets would be in the plane of the Sun's equator. Actually, the Earth's orbit is inclined to the equatorial plane of the Sun by a full seven degrees, and the path of Mercury is tilted by another seven degrees with respect to that of the Earth. But there is another difficulty which is much more serious.

If we examine one body revolving round another, and consider together its mass, its distance and its velocity, we arrive at what is known as angular momentum. It is a fundamental principle that angular momentum can be transferred, but never destroyed, so that on Laplace's theory all the angular momentum now possessed by the Sun and the planets must originally have been contained in the gas-cloud. At present, almost all the angular momentum of the Solar System resides in the four giant planets Jupiter, Saturn, Uranus and Neptune—but on the Nebular Hypothesis we would expect to find most of the angular momentum concentrated in the Sun itself. This would mean that the Sun would rotate fairly quickly. Actually it is a slow spinner, and takes over 25 Earth-days to complete one turn on its axis.

By the end of the last century, mathematicians had launched so many savage attacks on the Nebular Hypothesis that it had been cast on to the scientific scrap-heap. Next came a crop of theories involving both the Sun and a passing star. The first of these 'tidal theories' was outlined by Chamberlin and Moulton, in America, in 1900.

Space is very sparsely populated. The Sun, which is a normal star, has a diameter of over 860,000 miles; but if we represent it by a tennis-ball, its nearest neighbours will be over a thousand miles away. This means that collisions between two stars must be very rare indeed, at least in our part of the universe. Two gnats flying about inside the Royal Festival Hall would be unlikely to meet head-on, but the chances are much greater than that of a stellar collision.

On the other hand, such a catastrophe cannot be ruled out; and on the Chamberlin–Moulton theory, a wandering star approached the Sun so closely that tremendous gravitational strains were set up. Matter was pulled off the Sun, and remained as a cloud after the intruder had done its worst and had receded into the distance. The material in the cloud began to condense into small bodies or 'planetesimals'; these in turn combined to make larger planets, and the final result was the Solar System we know today.

Once a planetesimal had reached a diameter of 100 miles or so, its own gravitational attraction would become strong enough for it to collect extra material, and so it would grow relatively quickly. Unfortunately for the theory, it has been shown that under the conditions assumed by Chamberlin and Moulton no aggregation could ever reach sufficient size and mass for the collecting process to begin. The high temperature of the material torn away from the Sun would make the gases disperse long before they could condense into planets.

Sir James Jeans, remembered today as much for his popular books and broadcasts as for his theoretical researches, recognized this difficulty, and worked out a modified tidal theory according to which the planets condensed from a long, cigar-shaped filament drawn out of the Sun by the pull of the passing star (Fig. 3). Certainly this would explain why the largest planets (Jupiter and Saturn) are more or less in the middle of the Solar System, since they would have been produced from the fattest part of the 'cigar', with the smallest planets at the tapering ends. The same sort of process could account for the satellite families of the giant planets, with the Sun cast in the disruptive rôle played originally by the wandering star.

Again the mathematicians swarmed to the attack. In an effort to improve

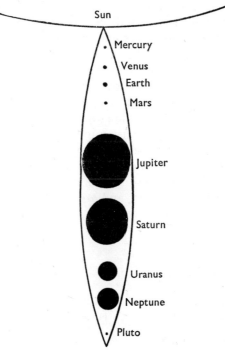

Fig. 3. Jeans' theory of the formation of the planets. The theory has now been discarded.

matters, Sir Harold Jeffreys suggested that the intruding star actually struck the Sun a glancing blow. The idea was not entirely new—a somewhat garbled version of it had been published earlier by A. W. Bickerton—but at first it looked promising. Then it, too, was found to have serious weaknesses, since the actual mode of planet formation remained much the same as on the Jeans' pattern. It is significant, too, that the planetary orbits are almost circular (neglecting Pluto, which seems to be a special case). On any collision theory they would be expected to be highly elliptical.

Quite apart from these objections, it now seems more than doubtful whether planets could be the result of any tidal or collision process. We must start again, and look elsewhere for a solution to our problem.

The Sun, of course, is a solitary star, but stellar pairs are very common in the universe. They are known as binary systems, and are of various kinds. Sometimes the two components are perfect twins; sometimes one member of the pair is decidedly brighter and more massive than the other; sometimes the components are strikingly unequal.

Probably the most famous of all the binaries is Mizar (Fig. 4), the second star in the 'tail' of the Great Bear (or, if you like, the handle of the Plough). Mizar is easily identifiable with the naked eye, because a much fainter star, Alcor, lies close beside it. When a telescope is used, Mizar itself appears as two stars, so close together that to the unaided eye they appear as one. This is no mere line of sight effect; the components are genuinely associated, and are revolving round their common centre

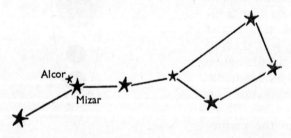

Fig. 4. Ursa Major (the Great Bear), showing the binary star Mizar with its more distant companion Alcor.

of gravity much as the two bells of a dumb-bell will do when twisted by their joining arm. Sirius, the brightest star in the sky, also has a binary companion, but small telescopes will not show it, since it has only 1/10,000 of the luminosity of its brilliant primary.

Attempts have been made to explain the origin of the planets by supposing that in the remote past the Sun itself was one component of a binary system. For instance, H. N. Russell believed that it was the companion star which was struck by the intruder, causing enough débris to account for the planets; R. A. Lyttleton considered that a near approach by the wandering star would be enough to wrench the binary companion away from the Sun, planet-forming material being scattered in the process. F. Hoyle, rather later, dispensed with the intruding star altogether, and suggested that the Sun's erstwhile companion exploded, shedding much of its material in the solar neighbourhood before departing permanently. Certainly we do occasionally see stellar outbursts of this kind, and we call them supernovæ,* but the theory is impossible to check, and it has not met with much support. Hoyle himself has now discarded it.

Oddly enough, we know rather more about the origin of the Sun than about that of the Earth. There is little doubt that the stars begin their careers by condensing out of nebulæ, which are clouds of dust and gas in space. Many nebulæ are known, and some of them, notably the Great Nebula in Orion's sword, are visible with the naked eye. It is generally thought that the two components of a binary system are born separately, relatively close to each other inside the nebula, instead of being the result of the break-up of an originally single star.

This has led on to another interesting theory, due to G. P. Kuiper in the United States. Kuiper regards the Solar System as a degenerate binary in which the second mass did not condense into a single body, but was spread out. Instead of two stars, the initial result was one star (the Sun) attended by a large number of condensations or 'protoplanets'. The total mass of the protoplanets would amount to about one-tenth of the mass of the Sun, which is reasonable enough. Once formed,

---

* The last supernova in our own star-system or Galaxy was seen as long ago as 1604, but many have been recorded in external galaxies.

the protoplanets would contract to form the planets we know, drawing in other material from around, and leaving most of the rest of the solar cloud to be lost to interstellar space. Here again we have an idea which is logical and plausible, but is extremely difficult to check; we have no proof that the Sun ever had a binary companion, degenerate or otherwise.

More recently still there has been wide support for theories of what may be called the 'old-fashioned' type, inasmuch as they mark something of a reversion to Laplace's Nebular Hypothesis, albeit in vastly improved form. During the 1950s Otto Schmidt in Russia and Carl von Weizsäcker in Germany put forward theories which are essentially similar, though different in detail. This time it is supposed that the Sun used to be surrounded by a huge 'envelope' of material, made up chiefly of the two lightest elements, hydrogen and helium. Collisions and friction between the particles in the cloud led to the formation of a circular, disk-shaped shell. As time passed, aggregations grew up; when they became massive enough they drew in more material, so that the planets were produced. Once again we have an accretion process, but we can thankfully dispense with the intruding star of Jeans and Jeffreys and the impossible gaseous rings of Laplace.

Of course, we have to explain how the Sun came to collect its 'envelope' in the first place. One suggestion is that it passed through a nebula, and came out surrounded by a vast, tenuous cloud. Alternatively, it is possible that the Solar System began as a simple gas-and-dust cloud, so that as shrinking took place, under the influence of gravity, the planet-forming material was 'left behind', so to speak; obviously this brings even closer to Laplace! Hannes Alfvén, of Sweden, believes that magnetic forces played a major rôle in the building-up of planets from the solar cloud, and most authorities tend to agree with him. In any case, the Sun represents the central remnant of the original cloud.

According to the new outlook, planetary systems are likely to be common in the universe, and other stars may well have retinues of their own, whereas on the older Jeans–Jeffreys theories the Solar System would be something of a cosmical freak. I shall return to this point in Chapter 16, because it is particularly significant.

Though we are still uncertain of our facts, we are slightly more confident than we used to be a couple of decades ago, and some sort of accretion process from a solar cloud appears to be the most logical answer. No doubt we shall learn more from the probes which are now being planned. Meanwhile, what can we discover about the future of the Solar System?

We are utterly dependent upon the Sun. Life on Earth can exist only because the conditions here are exactly right for it; if the Sun became more luminous our oceans would boil and our atmosphere would be stripped away, while if the supply of solar energy fell off to any marked degree we should be promptly and unceremoniously frozen. Fortunately for us the Sun is a stable, well-behaved star. It consists largely of hydrogen, and it shines because near its core the hydrogen is being transformed into another element, helium; as this happens, energy is released, and mass is lost. As we noted earlier, the Sun is losing four million tons every second, but it is so massive that it will not change perceptibly for several thousands of millions of years to come. Very slight fluctuations in its output can do no more than cause periodical Ice Ages and warmer spells, as has happened in the past all through the Earth's history.*

Yet the Sun cannot survive for ever. Eventually its supply of hydrogen 'fuel' will run low, and the Sun will change radically, with results which are certain to be disastrous as far as the Earth is concerned. It used to be thought that there would be a gradual, remorseless cooling-down, so that life on our world would be wiped out by creeping cold. Nowadays, it is believed that when the Sun can no longer maintain itself by the hydrogen-into-helium conversion, its core will shrink, while its outer layers will expand; the Sun will become a star of the type known as a Red Giant, and for a while it will send out much more energy than it does at present. Even if the Earth survives, life here will not. Subsequently the Sun will collapse into a small, massive body well on the road to its own death; we may assume that it will reach its final condition as a cold, dark globe, lightless and heatless, still circled by its remaining

---

* The cause of the Ice Ages is a matter for debate, but nowadays most authorities blame minor fluctuations in the Sun. The last cold spell ended 10,000 years ago.

planets. But this will not concern humanity, for we will not be there to see. In the remote past, the Sun was responsible for the creation of the Earth; in the end, it must inevitably destroy us.

Chapter Three

THE MOVEMENTS OF THE PLANETS

To the casual observer who has no knowledge of astronomy, a planet looks very much like a star. True, each planet has its own special characteristics; Venus and Jupiter stand out because they are so bright, while Mars is strongly red. Yet Saturn, in particular, is remarkably starlike. It is not even true to say, as many books do, that stars twinkle and planets do not. Since a planet shows up as a tiny disk rather than a point source of light, it does not twinkle to any marked extent when high up, but when low over the horizon the effect can become very noticeable. (Twinkling, of course, is due entirely to the Earth's unsteady atmosphere, so that it is least for objects near the zenith or overhead point.)

The only real reason why the pre-telescopic observers could single out the planets was because of their motion against the starry background. On the other hand, they firmly believed that everything in the sky, including the planets, moved round the Earth. This scheme of things was brought to its highest perfection by Ptolemy of Alexandria, who lived between A.D. 120 and 180. His system is always known as the Ptolemaic, though

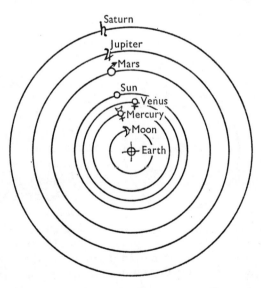

Fig. 5. The Ptolemaic theory, according to which the planets moved in perfect circles round a stationary Earth.

to be strictly accurate Ptolemy himself did not invent it.

On the Ptolemaic theory (Fig. 5), the Earth occupied the central position. Round it, in perfect circles, moved the Moon, Mercury, Venus, the Sun and the more remote planets, Mars, Jupiter and Saturn. Beyond Saturn came the sphere of the fixed stars. All celestial orbits were perfectly circular, and the heavens turned round the Earth once a day.

The main drawback was that the movements of the planets in the sky did not fit in with any simple, uncomplicated theory. For instance Mars, Jupiter and Saturn sometimes perform slow 'loops', and there are times when they reverse their direction of motion, so that they shift from east to west against the stars instead of west to east. Ptolemy, who was an excellent observer as well as being a brilliant mathematician, knew this quite well. Moreover he was in possession of a good star catalogue, and his measurements of planetary positions were extremely good in view of the fact that they had to be made without the help of a telescope.

Some kind of solution had to be found, and unfortunately Ptolemy selected the wrong one. Rather than abandon his perfect circles—after all, the circle was the 'perfect' form!—he believed that each planet must move in a small circle or epicycle, the centre of which (the deferent) itself moves round the Earth in a circle (Fig. 6). Even this would not do, and there was no choice but to introduce more and more epicycles, as well as other refinements. Finally the scheme was made to fit the apparent motions of the planets reasonably well, but it had become hopelessly clumsy and artificial.

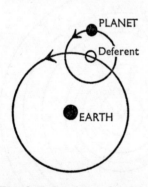

Fig. 6. Epicycles and deferents. On the Ptolemaic theory, a planet moved in a small circle or epicycle, the centre of which (the deferent) itself moved round the Earth in a perfect circle.

Ptolemy's ideas were accepted with almost no argument for more than a thousand years after his death. Even in the Middle Ages, any proposal to dethrone the Earth from its supreme central position was regarded as blasphemous. Then, in the mid-sixteenth century, a Polish canon, Copernicus, took a long, hard look at the whole situation, and published a book which led eventually to the overthrow of the Ptolemaic system. The book was called *De Revolutionibus Orbium Cœlestium* (Concerning the Revolutions of the Celestial Bodies), and it proved to be a landmark in scientific history. It also led to a great deal of unpleasantness.

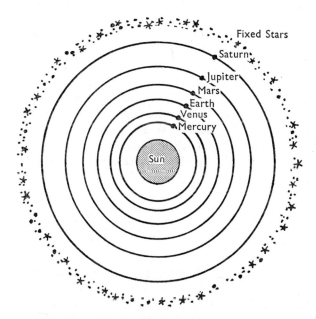

Fig. 7. The Copernican system, according to which the planets moved in circular orbits around the central Sun.

Let us admit that Copernicus was by no means correct in all he said. He still kept to the idea of perfectly circular orbits, and he was even reduced to bringing back Ptolemy's epicycles. Yet he made the one great fundamental advance which was needed; he removed the Earth from the centre of the Solar System, and put the Sun there instead (Fig. 7). From being the most important body in the universe, our world was relegated to the status of an ordinary planet.

This sort of heresy did not appeal to the Christian Church,

which was inclined to take drastic action against those who dis-
agreed with it. Copernicus was in Holy Orders, and he was
keenly aware of the danger of persecution. He avoided trouble
by the simple means of withholding publication of his book
until he was at the very end of his life, but some of his later
supporters were less prudent. In 1600 Giordano Bruno was
burned at the stake in Rome, one of his crimes being that he
taught the Copernican theory rather than the Ptolemaic.

The scientific revolution had begun, but it took a hundred
years to complete. Oddly enough, the man responsible for the
next major advance—the Danish astronomer Tycho Brahe—
was no supporter of Copernicus; he stoutly maintained that the
Sun must move round the Earth, and nothing would shake his
belief. However, he made a great many accurate observations
of star positions, and also measured the movements of the
planets, particularly Mars. When he died, in 1601, all his
results came into the hands of his last assistant, a young German
named Johannes Kepler. Kepler used them well—but with a
result quite different from anything that Tycho could have
expected.

Kepler started with the assumption that the Sun lies in the
centre of the Solar System. He then did his best to work out a
theory which would suit Tycho's observations. He concentrated upon Mars, and toiled away for several years, only to
meet with failure after failure. The observations nearly fitted,
but not quite. Either Tycho's work was inaccurate, or else
there was something badly wrong with the theory.

Kepler had implicit faith in Tycho as an observer, and at
last he found the cause of the trouble. The planets do indeed
move round the Sun, but they do so in orbits which are elliptical rather than circular.

One way to draw an ellipse is to stick two pins into a board,
an inch or two apart, and join them with a thread, leaving a
certain amount of slack. Now draw the thread tight with a
pencil, and trace a curve. The result will be an ellipse, the two
pins marking the *foci*. If the pins are wider apart, with the same
length of thread, the ellipse will be longer and narrower. The
distance between the foci is thus a measure of the eccentricity
of the ellipse.

In the case of the Earth, the Sun occupies one focus of the

ellipse, while the other focus is empty.* However, the shape of the orbit is quite unlike the eccentric ellipse shown in the left-hand diagram. The Earth's path is nearly circular, so that our distance from the Sun never ranges beyond 1½ million miles from the average value of approximately 93,000,000 miles. In the right-hand diagram (Fig. 8), I have drawn the orbit to the correct eccentricity—and you will agree that it looks very circular! Yet it was slight departures from the circular form which led Kepler on to his great discovery. It was fortunate for him that he was so confident of Tycho's observing ability.

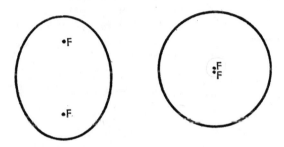

Fig. 8. Ellipses. (*Left*) An eccentric ellipse, such as the orbit of a typical comet. (*Right*) An ellipse which is almost circular, such as the orbit of the Earth.

Mars has an orbit which is rather more eccentric than ours; its distance from the Sun ranges between 128¼ million miles at its closest or *perihelion* out to 154½ million miles at its farthest or *aphelion*. If the path of Mars had been as nearly circular as that of (say) Venus, it would have taken Kepler much longer to find the answer to his problem, if indeed he would have been able to do so at all.

Once he had taken the decisive step, Kepler was able to draw up three fundamental Laws of Planetary Motion. The first states that a planet moves round the Sun in an ellipse, with the Sun at one focus. The second states that the *radius vector* sweeps out equal areas in equal times, while the third gives a

* This is another over-simplification, because we have to reckon not only with the Sun-Earth pair but also with all the other planets. However, the general scheme should be clear enough.

relationship between a planet's time of revolution *(sidereal period)* and its distance from the Sun.

The second law is worth a little further explanation, because it has a particular significance. The radius vector is the line joining the centre of the planet to the centre of the Sun. In the diagram (Fig. 9), the shaded area is equal to the dotted area

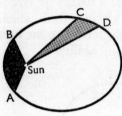

Fig. 9. Kepler's Second Law. The blacked area B–Sun–A must be equal to the shaded area C–Sun–D, assuming that the planet moves from A to B in the same time that it takes to move from C to D.

—it being assumed that the planet takes the same time to move from A to B as it does from C to D. In other words, a planet moves at its quickest when near perihelion. It follows that the nearer planets are the faster-moving. Mercury races on at a mean rate of 30 miles per second; the Earth at $18\frac{1}{2}$* and Jupiter at 8, while Pluto, the tortoise of the Solar System, has an average velocity of only 3 miles per second.

Two of the planets, Mercury and Venus, are closer to the Sun than we are, so that they have their own way of behaving; they are sometimes termed the 'inferior planets'. To begin with, we can never see them against a dark background except when they are low down in the sky. They always lie in roughly the same direction as the Sun, and are at their brightest when in the west after sunset or in the east before dawn. Moreover, they show phases similar to those of the Moon. The diagram here (Fig. 10), shows what happens.

Let us consider Mercury first. When at position 1, it has its dark side turned toward us, and is 'new', so that it cannot be seen at all. This is known as *inferior conjunction*. As it moves along in its orbit, a little of the daylight side begins to be turned in our direction, so that Mercury comes into view. It shows up as a crescent; then, at position 2, it is a half-disk *(dichotomy)*. It then becomes *gibbous*, or three-quarter phase, and becomes full at position 3. At full, or *superior conjunction*, it is almost behind the Sun in the sky, so that it is difficult to see even with a telescope. After superior conjunction, the phases are repeated in the re-

* $18\frac{1}{2}$ miles per second = 66,000 m.p.h.

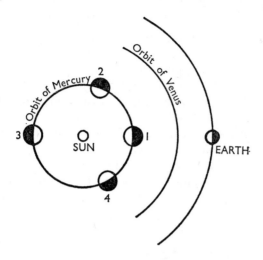

Fig. 10. Phases of Mercury. (1) New.
(2) Dichotomy (half-phase). (3) Full.
(4) Dichotomy. For the sake of clarity, the
Earth's movement round the sun is not
taken into account in this diagram.

verse order: gibbous, half at position 4, crescent, and then back
to new. When an evening star, Mercury is on the wane; when
a morning star, it is waxing. Because of its quick motion, Mercury passes through all its phases several times a year; in 1971,
for instance, it was new on April 19, August 26 and again on
December 12. Of course, the Earth's own movement round the
Sun must be borne in mind; for the sake of clarity, I have disregarded it in the diagram.

The angular separation between Mercury and the Sun in the
sky is never as much as 30 degrees, and it never rises long before
the Sun or sets long after it. Also, it is a small world, not a great
deal larger than the Moon; and as the phase increases, the
apparent diameter shrinks, because Mercury is drawing away
from the Earth. City-dwellers will probably never see it with
the naked eye, though people who live in the country may see
it shining as a fairly bright, starlike point when it is best placed.

At some inferior conjunctions Mercury passes directly
between the Sun and the Earth, so that for a few hours it may
be seen as a black spot in transit against the brilliant solar disk.

This last happened in May 1970, and will do so again in November 1973. Transits do not occur at every inferior conjunction because Mercury's orbit is tilted to that of the Earth, and usually the lining-up is not exact, allowing Mercury to pass unseen either above or below the Sun in the sky.

Venus behaves in the same way as Mercury, but is much more conspicuous, partly because it is farther from the Sun and closer to the Earth, and partly because it is much larger and more reflective than Mercury. Transits are very rare, and the next is not due until the year 2004.

The remaining planets lie beyond the orbit of the Earth in the Solar System, and are therefore more convenient to observe. Mars is shown in the diagram (Fig. 11), but again I have disregarded the orbital eccentricity—which, as we have noted, is more marked than in the case of Earth.

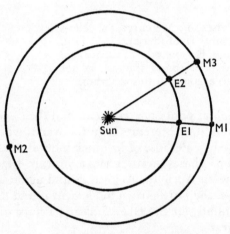

Fig. 11. Movements of Mars. With the Earth at E1 and Mars at M1, Mars is at opposition. A year later the Earth is back at E1, but Mars has reached only M2. The next opposition does not occur until the Earth has 'caught Mars up' (E2 and M3).

Obviously, no planet outside the Earth's orbit can ever pass through inferior conjunction, since it can never go between the Sun and the Earth. At the corresponding position in its orbit, Mars, at M1, is certainly lined up with the Sun and the Earth; but this time the Earth is in the mid-position, with Mars opposite to the Sun in our sky, and best placed for observation. This is termed *opposition*. With Mars, the average interval between successive oppositions (the *synodic period*) is 780 days. To a Martian observer, the Earth would then be at inferior conjunction.

34

Like Mercury and Venus, Mars and the outer planets can reach superior conjunction. In this position, Mars is virtually behind the Sun as seen from Earth, so that it is above the horizon only in daylight, and for a few weeks it is to all intents and purposes out of view.

The reason why Mars comes to opposition only once in 780 days or so is because both it and the Earth are moving. The Earth takes approximately 365 days to make one circuit of the Sun; and if we start with the Earth at E_1 and Mars at M_1, the Earth will have returned to its original position after 365 days. Mars, however, has a longer 'year'—687 of our days—and it will not have arrived back at M_1. It will have travelled only as far as M_2. The Earth has to catch it up, which will happen with the Earth at E_2 and Mars at M_3. This means that oppositions of Mars do not occur every year; there was no opposition, for instance, in 1970.

Conditions are not quite the same for the more distant planets. They move so slowly compared with the Earth that they are much easier to catch up, so that they have shorter synodic periods, and come to opposition at intervals of little over a year. For example, Jupiter's synodic period is 399 days, so that the opposition of May 1971 is followed by another in June 1972.

We can now explain the periodical backwards or *retrograde* movements which so perplexed Ptolemy. In the diagram (Fig. 12), the orbits of Earth and Mars are shown, with the apparent motion of Mars in the sky. Between positions 3 and 5 the Earth is catching Mars up and passing it, so that for a while Mars seems to 'loop the loop'.

The orbits of the planets lie in very much the same plane, so that we are not far wrong in drawing them on a flat sheet of paper. This means that the planets can never wander near the celestial poles; they keep relatively close to the *ecliptic*, which is the plane of the Earth's orbit, and may also be defined as the apparently yearly path of the Sun against the stars. The ecliptic passes through the twelve constellations of the Zodiac: Aries (the Ram), Taurus (the Bull), Gemini (the Twins), Cancer (the Crab), Leo (the Lion), Virgo (the Virgin), Libra (the Balance), Scorpio or Scorpius (the Scorpion), Sagittarius (the Archer), Capricornus (the Sea-goat), Aquarius (the Water-bearer)

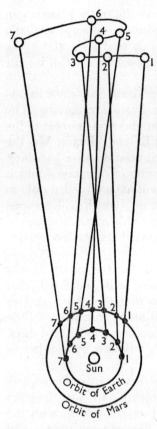

and Pisces (the Fishes), though a thirteenth group, Ophiuchus (the Serpent-bearer) intrudes into the Zodiac for some distance between Scorpio and Sagittarius. Of course, the coincidence is not exact —otherwise Mercury and Venus would appear in transit against the Sun at every inferior conjunction— but none of the planets, apart from Pluto and some of the asteroids, can leave the Zodiac.

A brief account of the motions of the planets can give little idea of the enormous problems which face the mathematical astronomer. Each planet pulls upon its fellows, producing disturbances or *perturbations* in the orbits; the satellites of the Earth and the giants also have to be taken into account, and there are any number of other complications, such as the 'relativity effect' predicted by Einstein—which has turned out to be quite important in the case of Mercury.

Fig. 12. Retrograde movement of Mars. As the Earth 'catches up' with Mars and passes it, the movement will seem to be in a retrograde direction, so that between positions 3 and 5 Mars will seem to move 'backwards' in the sky (east to west, instead of the usual west to east relative to the stars).

Astronomical handbooks publish tables of the planets, but it should be remembered that these are only for the mean orbits, and that a planet does not follow exactly the same track in each journey round the Sun. An orbit depends primarily on the velocity of the planet, and this velocity is not constant. Taking every predictable perturbation into account, it is possible to compute an *osculating* orbit, which is much more accurate. For instance, the mean value for the eccentricity of the orbit of Neptune is known with high precision, but in early 1961 the planet was moving according to an orbit whose eccentricity was slightly less than the mean. Drawn on a diagram which would fit on to this page, the difference between the two orbits would be concealed by the breadth of an inked line; but the actual difference between the hypothetical 'mean' position of Neptune at that time (calculated according to the mean orbit) and the real position of the planet (calculated according to the osculating orbit) amounted to no less than 9,000,000 miles.

This shows us how great are the complications with which theoreticians have to deal. Yet there is no doubt that they have succeeded wonderfully well in their attempts to unravel the problems of celestial mechanics. Uncertainties and errors are bound to remain, but our knowledge is much more complete than was the case not so very long ago. Otherwise, we could not hope to launch a successful rocket probe out to a planet far away in the Solar System.

Chapter Four

ROCKETS TO THE PLANETS

THERE ARE TWO ways of exploring the planets. One—the age-old method—is by telescope; by observing from the surface of the Earth we can learn a great deal. The other is by means of rocket probes. The first step was taken in 1959, when the Russians sent up their first vehicle toward the Moon. Since then we have had rockets to Venus and Mars, together, of course, with the lunar landings. Before 1980 we may confidently expect probes to have been sent out to the other planets as well, and there will probably have been several 'soft' landings on Mars.

In a book dealing principally with the planets themselves, I do not think that it is necessary to say much about rocket theory, because it is now so much a part of our everyday life (though this was far from being the case when the first edition of this book appeared, way back in the early 1950s!).* Therefore I will be brief; but something must be said, because even today there are some widespread misconceptions as to how a rocket actually works.

To begin with, it does not 'push against' anything except itself. A November the Fifth rocket consists of a hollow tube filled with gunpowder, plus a stick to give stability in flight. When the powder burns, it gives off hot gas; this gas rushes out of the exhaust, and propels the rocket skyward. It flies because of what Sir Isaac Newton so rightly called the principle of reaction: every action has an equal and opposite reaction. So long as the gas continues to stream out, the rocket will go on accelerating.

The essential point here is that the Earth's atmosphere is not concerned. Indeed, it is a hindrance, because it sets up resistance, and has to be pushed out of the way. Rockets are at their best in 'empty' space, where there is virtually no resistance at

---

* I still have a copy of the review of the first edition of this book, written by an eminent astronomer who shall remain nameless. 'Even though Mr. Moore's idea of sending rocket vehicles out to Mars and Venus is so obviously naïve . . .'

all. This is why rockets, and rockets alone, can be used for flight beyond the Earth.

The first man to realize this clearly was a Russian, Konstantin Eduardovich Tsiolkovskii, who published some papers about it seventy years ago. They met with no comment at all, partly because they appeared only in obscure Russian periodicals and partly because the whole idea of space-travel was officially regarded as crazy. (At about the same time, Orville and Wilbur Wright were making their first aeroplane flights; they too were ignored at first, though before long their work became too obtrusive to be overlooked.) Tsiolkovskii also realized that solid fuels, such as gunpowder, are of no use for sending vehicles to the Moon; they are too weak, and they are not controllable. Instead, Tsiolkovskii planned to use liquid propellants, and to replace the crude charge of gunpowder with a proper rocket motor. He was not a practical rocket experimenter—activities of such a kind would hardly have been welcomed in Tsarist Russia!—and it was only in 1926 that the first liquid-propellant rocket was sent up, not by Tsiolkovskii, but by an entirely independent American researcher, Robert Hutchings Goddard. During the 1930s more experiments were carried out, and then, during the war, the Germans built rockets powerful enough to bombard England. Many people have rueful memories of Hitler's V.2 weapons; yet although they were built for destruction, they were the direct ancestors of the space-craft of today.

They could not, of course, break free from the pull of the Earth, because they could not go fast enough. Here let us dispose of another popular misconception. No rocket can leave the Earth's gravitational field, which is theoretically infinite. Also, it is theoretically possible to travel from the Earth to the Moon in a vehicle which can reach no more than a few miles per hour; but you would have to go on using fuel all the way, which would be hopelessly uneconomical (apart from the fact that the journey would take a very long time indeed). The only practical method is to give the vehicle an initial acceleration which will break it free from the Earth's pull without further expenditure of fuel. There is some analogy, though not an accurate one, with a cyclist who pedals so hard that he can then coast up a hill without using any further energy.

The critical velocity is 7 miles per second, or around 25,000 m.p.h. If the rocket can accelerate to this velocity, it will not fall back to the ground; it will continue moving outward, and can 'coast' toward the Moon. Therefore, 7 miles per second is termed the Earth's escape velocity.

This was no new idea even in Tsiolkovskii's time. One of those who had grasped it was Jules Verne, the great French story-teller who wrote a novel about a lunar voyage as long ago as 1865. However, Verne planned to send up his travellers in a projectile fired from the muzzle of a huge gun. This involved a sudden departure at full escape velocity, and there are several reasons why this is out of the question. Any vehicle moving at such a speed through the dense lower layers of the Earth's atmosphere would be destroyed by the friction generated against the air-particles. Moreover, the shock would have unfortunate consequences for anyone who happened to be inside the projectile!

With a rocket probe, the acceleration can be gradual, so that the vehicle reaches escape velocity only when it has passed beyond the thick part of the Earth's atmosphere. The amount of propellant needed is colossal by any standards, and for this reason a rocket launcher is built in several 'steps'. Immediately after blast-off, the large lower motor* does all the work; when it has used up all the propellant it can carry, the whole of the bottom stage of the launcher separates and falls back to the ground. The motors in the second stage are fired, and when they too are exhausted the motors in the third stage take over. Only the upper section of the vehicle goes the whole way. With the Apollo Moon programme, the pre-launch vehicle stands over 360 feet in height—equal to St. Paul's. Yet only a 22-feet cone, carrying the astronauts, comes back intact at the end of the journey.

The Moon is a special case. It is our companion, and it always stays with us in our never-ending journey round the Sun. A trip there and back takes less than a fortnight. When we come to consider the planets, we are faced with much greater problems, because the distances involved are increased, and also because the planets do not stay obligingly close to us.

* Or, rather, motors. The design of a space-probe launching vehicle is almost incredibly complicated.

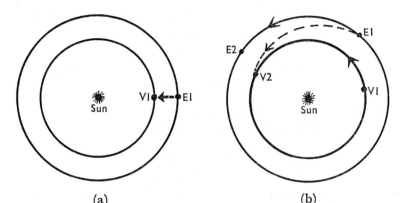

|(a)|(b)|

Fig. 13. Orbit of a Venus probe (*a*) Impracticable path, from the Earth (E1) straight to Venus (V1) (*b*) Transfer orbit. The probe is launched with the Earth at E1 and Venus is at V1., The probe swings inward, and meets Venus at V2. Meanwhile the Earth has moved on to E2.

Since Venus was the first planet to be selected as a target, we will consider it first; and the diagrams will help. In the left-hand diagram (Fig. 13a) Venus is at V1 and the Earth at E1; the minimum distance between the two worlds may be less than 25 million miles, which is roughly a hundred times as far away from us as the Moon. Why not wait for a suitable moment, with Venus at inferior conjunction, and then fire a probe straight across the gap?

Alas, this cannot be done. Without continuous application of power, no vehicle can move in a path of this kind, and the amount of propellant needed would be prohibitive. Moreover, both Venus and the Earth are moving, and no space-craft can manœuvre in the same way as an aircraft.

The procedure actually followed is very different. Basically, what is done is to take the probe up in a step-vehicle, and then slow it down relative to the Earth. Remember that once a vehicle is in space, and unpowered, it moves in exactly the same way as a natural body would do, and it obeys Kepler's Laws. When the probe has been 'slowed', and the motors switched off, it will not stay with the Earth; it will begin to swing inward toward the Sun, picking up velocity again as it does so. Eventually it will reach the orbit of Venus. If all the calculations have been correct, it will rendezvous with Venus as shown in the right-hand diagram (Fig. 13b). If the launching has been faulty, and Venus is not there to meet the probe, then the inward swing will continue until the probe has picked up enough velocity to start moving outward again; and it will continue moving round the Sun in an elliptical orbit until it collides with some solid body and is destroyed—a misadventure which may well not happen for many millions of years, if at all.

Let us suppose that everything has gone well. (Mid-course corrections can be, and are, made by commands sent out from Earth.) Firing the motors in the probe itself can then result in a landing on Venus, as has now happened several times. Otherwise, the vehicle will swing past Venus, and then continue in a modified elliptical orbit.

Quite clearly, most of the journey from the orbit of the Earth to that of Venus is done in 'free fall'—or coasting, if you like—which is the only possible method, because of the limited amount of propellant that can be carried. When nuclear fuels become available, things will be different; but as yet we have to rely upon liquids, which are not nearly so powerful as the rocket designers would like.

There would be little sense in sending a probe out to Venus if it could tell us nothing. However, the vehicles carry a surprising amount of equipment, all made to fit into as small a compartment as possible. Since 1962, our whole knowledge of Venus has been transformed by the American and Russian probes which have been sent there.

The diagrams show that in order to reach Venus by the free-fall method, the probe has to move in a curve for a considerable way round the Sun, and this takes time. Mariner 2, the first successful Venus explorer, was launched on August 27, 1962, and did not reach its point of closest approach to the planet until December 14. The travel-times of the subsequent probes have been of the same order: around four months. There is nothing that can be done about this until better propellants are available, so that for the moment we must be content with unmanned vehicles.

Now let us turn to Mars, where the problems are of the same basic type. This time the probe has to be speeded up relative to the Earth, so that it swings outward towards the orbit of

Mars, losing speed steadily. If it does not land (which can be achieved only by using the built-in motors, fired by remote control from Earth) it too will continue indefinitely in an elliptical orbit round the Sun after it has by-passed Mars. With Mars, too, the first success was American. Mariner 4 was launched on November 28, 1964, and reached the neighbourhood of Mars on July 15, 1965, over seven months later.

With the more remote planets, the times of travel become ominously long, but there is one way of cutting them. This is by the method sometimes known by the irreverent name of 'interplanetary billiards'. Basically, it involves using the gravitational pull of one planet to accelerate a probe on toward the next.

Jupiter, the largest and most massive planet in the Sun's family, is the key to what is often called the Grand Tour. When a probe reaches the neighbourhood of Jupiter, it can be made to swing round it, picking up extra velocity from the tremendous gravitational field. At the moment there are plans to send up a vehicle which will be launched in 1977, and will approach Jupiter by 1981 (Fig. 14). After being accelerated, it will con-tinue on its curving path outward, until by 1985 it will have reached the region of Uranus. Again there will be a swing-round, and an accelera-tion, so that by 1988 the probe will have arrived near Neptune. Another Grand Tour is scheduled to begin in 1979; this time the targets will be Jupiter, Saturn and Pluto, in that order.

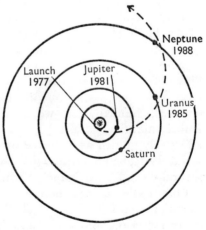

Everything depends upon the planets being in the right place at the right time, and it so happens

Fig. 14. One variant of the Grand Tour: Earth–Jupiter–Uranus–Neptune.

that the end of the 1970s will be ideal in this respect. If we miss the opportunity, we must wait for 200 years before we can try another Jupiter–Uranus–Neptune trip, and over 100 years before Jupiter is again suitably placed to help us on our way to both Saturn and Pluto. Of course, this does not mean that no long-range journeys can be undertaken in the meantime; but they will have to be carried out without the 'billiards' technique', so that the times of travel will be much longer. (I am assuming, of course, that we will still have to depend upon our 1970–type liquid propellants, which is no doubt a very pessimistic outlook.)

If unlimited funds were available, no doubt a whole series of Grand Tour type probes would be launched before 1980. The scope is very great indeed. For instance, it has been suggested that the main vehicle might dispatch smaller probes on to each planet as it came within range; it might even be possible to establish vehicles in closed orbits round Jupiter or one of the other giants. Unfortunately the cost would be extremely high, and with the present cut-back in funds it may be that the scientists will have to be content with something rather less ambitious. However, even a straightforward vehicle which was able to send back close-range photographs and fundamental data would represent an extremely important advance.

There is also considerable uncertainty about the Russian plans. Undoubtedly the Soviet workers are very much concerned with long-range planetary exploration with unmanned vehicles; they seem to have overcome the communications difficulties which held them back in the early 1960s, and they have, of course, had a magnificent triumph with their 'Moon crawler', Lunokhod I, which was sent to the lunar Mare Imbrium at the end of 1970, and proved to be capable of being effectively guided as it moved around on the lunar surface. It would be logical to assume that the Russians, also, will attempt several Grand Tour launches while the conditions are right.

One trouble is that there is no time for proper testing. The essential need is to have a vehicle which can keep in working order for years, and can maintain radio contact. The Grand Tours must be attempted without rehearsals; one can only hope that they will be successful.

Incidentally, the same sort of technique can be used for the inner regions of the Solar System. According to current plans, the first probe to Mercury will go there by way of Venus, and will make use of Venus' gravitational field.

When nuclear rockets have been perfected, the times of travel to remote worlds will be cut considerably. Not only will greater speeds be possible, but the probes will be able to go by shorter routes rather than making almost the whole of the journey under conditions of free fall. When the nuclear rocket comes into its own, the Grand Tour techniques will become less important; but at the moment, they seem to hold out our best hopes of obtaining close-range information from the giant planets.

Ideas of this sort do not sound in the least far-fetched today, incredible though they would have seemed only a decade or two ago. Events are moving so quickly that it is difficult to forecast what will happen in the next few years. Official American opinion states that the first men will reach Mars round 1990—though this must surely depend upon the development of propellants much better than those which took the Apollo astronauts to the Moon.

I am well aware that this account has been over-simplified, but to delve further into rocket techniques would be out of place here, and I hope that I have at least given a general indication of what is going on. Certainly the probes have caused a complete change in our outlook. We know much more about the planets now than we used to do before 1968, when Mariner 2 began its epic journey to Venus. Many of the facts we have learned have disappointed people who are anxious to believe in life beyond the Earth; but science is unrelenting, and for the first time we are confident that we are on the right track.

Chapter Five

MERCURY

OF ALL THE naked-eye planets, Mercury is much the least obtrusive. There must be many people who have never seen it, even though it is surprisingly bright when at its best. There is a story—probably untrue—that Copernicus never managed to find Mercury, because of mists rising from the river near his home. People who live in modern cities or industrial areas have no hope at all.

Yet Mercury is actually more brilliant than any of the 'fixed stars' apart from Sirius. The trouble is that the planet never appears against a dark sky. It seems to stay inconveniently close to the Sun, so that it is a naked-eye object only when very low in the west after sunset or very low in the east before sunrise. One has to glimpse it at the right moment, and it does not remain on view for more than an hour or so per night. From my home at Selsey, on the Sussex coast, I have found that the average number of nights when Mercury is visible with the naked eye is eighteen per year; but Sussex has clearer skies than most other parts of England, and I also have a sea horizon.

From more southerly countries Mercury is easier to see, and the ancients knew it well. Originally they thought that the 'evening Mercury' and 'morning Mercury' must be different bodies, but before long it was realized that the two were identical. On the Ptolemaic theory, it was the closest body to Earth apart from the Moon. Because of its quick movements it was named in honour of Hermes, the Messenger of the Gods. The Latin equivalent of Hermes is Mercury.*

At its mean distance of 36,000,000 miles, Mercury is the closest planet to the Sun. One known asteroid (Icarus) can swing still nearer, and can approach Mercury to within a few million miles; but otherwise only the comets have the temerity to move in those torrid regions. A century ago, however, an intra-Mercurian planet was believed to exist; its presence was

* The 'geography' of Mercury is still termed 'hermography', and we refer to hermocentric longitudes and latitudes.

regarded as well-established, and it was even given a name—Vulcan, after the blacksmith of the gods. The story of this so-called 'discovery' provides an excellent case of how even the most brilliant scientists can be misled.

The Director of the Paris Observatory at that time was named Urbain Le Verrier. He was probably the greatest astronomer of his day, and had achieved lasting fame by his success in tracking down the planet Neptune, about which I shall have more to say in Chapter 14. In 1860, some fifteen years after this triumph, Le Verrier studied the movements of Mercury, and found that something was wrong with his predictions. The 'messenger of the gods' was not quite where he should have been, and Le Verrier decided that Mercury was being pulled out of position by the gravitational force of some unknown planet.

Just as Le Verrier finished his calculations, there came a message from a French country doctor, Lescarbault, who claimed that he had watched an intra-Mercurian planet passing in transit across the face of the Sun. Remember that at some inferior conjunctions both Mercury and Venus show up in transit, and presumably any inner planet would do the same—which, incidentally, would be almost the only hope of observing it at all. Le Verrier made haste to visit Lescarbault. The interview must have been a curious one. Le Verrier had the reputation of being one of the rudest men in France, and Lescarbault was hardly a professional scientist; for instance, he recorded time by an old watch which had lost its second hand, and he wrote his observations down with chalk upon wooden boards, planing them off when he had no further use for them. Yet in spite of all this, Le Verrier decided that a new planet had genuinely been found. He named it Vulcan, and worked out that it must be 13,000,000 miles from the Sun, with a sidereal period of $19\frac{3}{4}$ days and a diameter of about 1,000 miles. He also calculated the times of future transits.

Unfortunately Vulcan has never been seen since, and it is now certain that what Lescarbault recorded was not a planet. It may have been a sunspot, but it is worth noting that Liais, from Brazil, had been observing the Sun at the exact time of Vulcan's supposed transit, and had seen nothing at all.

Interest was rekindled briefly in 1878, when there was a total eclipse of the Sun. At a total eclipse the Moon passes directly in front of the Sun, and blots out the bright disk, so that for a few minutes stars can be seen with the naked eye in the middle of the day. Two American observers, Watson and Swift, carried out a careful search, and claimed that they had recorded various unidentified objects. Alas, Watson's and Swift's observations agreed neither with the predicted Vulcan nor with each other, and they must certainly have been mistaken. Since then, the irregularities in the movement of Mercury have been cleared up without the need to bring in an unknown planet; Einstein's relativity theory has provided a full explanation, and has shown that Vulcan does not exist. Let us return, then, to Mercury.

Mercury is hard to study in detail partly because of its nearness to the Sun, and partly because it is small and comparatively remote (Fig. 15); it never comes much within fifty million miles of us. According to the French astronomers H. Camichel and P. Muller, its diameter is 2,900 miles. Other authorities increase this to 3,100 miles, but in any case Mercury is comparable with the Moon, though it is considerably more massive; its density is about the same as that of the Earth, and it has

Earth   Mercury

Fig. 15. Mercury and Earth compared.

0·06 the mass of our world. Go to Mercury, and you will have a mere 38 per cent of your Earth weight.

To make matters worse, Mercury is invisible when at its closest to the Earth. It is then at inferior conjunction, and its non-luminous night side is turned toward us. As the phase grows, the apparent diameter shrinks. All in all, it is not surprising that our knowledge of the surface features of Mercury remains rather slight, though the probes due to be launched within the next few years ought to put a very different complexion upon matters.

On the other hand, we are certain that the surface markings, vague and elusive though they may be, are real. When we look at Mercury we are seeing a hard, presumably rocky surface, unveiled by any atmospheric cloud or haze. Recent results indicate that Mercury has virtually no atmosphere at all—a point which must, I think, be discussed more fully.

The power of a planet to hold down an atmosphere depends upon two factors: the temperature, and the escape velocity. The Earth, with its equable climate and its escape velocity of 7 miles per second, can hold down oxygen, nitrogen and the other atmospheric gases with no difficulty at all. The Moon, where the escape velocity is a mere $1\frac{1}{2}$ miles per second, has been quite unequal to the task. Any atmosphere it may once have had has long since leaked away into space, because its particles were able to travel outward at a rate greater than the lunar escape velocity. Mars, with its escape velocity of just over 3 miles per second and its chilly climate, has managed to retain a thin atmosphere. Mercury—2·6 miles per second—seems to be just below the critical limit. With high temperature

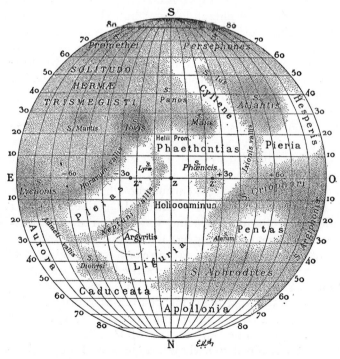

Fig. 16. Antoniadi's Map of Mercury.

the particles in an atmosphere can travel at increased speeds, so that they have a better chance of getting away; and Mercury, of course, is very hot indeed.

All this sounds logical enough, but there are some curious facts to be taken into account. Between 1924 and 1953 a long, careful study of Mercury was carried out by the Greek astronomer E. M. Antoniadi, who spent most of his life in France and observed with the aid of the magnificent 33-inch refracting telescope at the Observatory of Meudon, outside Paris. It was Antoniadi who produced the first really useful chart of Mercury (Fig. 16); I doubt whether it has been surpassed even yet. He was a splendid, accurate worker, and the 33-inch refractor could hardly be bettered, as I know from personal experience with it. In his classic monograph about Mercury, Antoniadi referred to atmospheric 'veils' which he regarded as more frequent and denser than those of Mars.

This is very strange. There is no doubt whatsoever about the clouds in the atmosphere of Mars; but Mercury is a very different kind of world. True, a very tenuous atmosphere would be enough to hold clouds of fine dust in suspension, but how would the dust get there in the first place? Winds would hardly suffice; the only answer would be volcanic activity, which on a planet such as Mercury does not sound very plausible.

Antoniadi's reputation was so great that until recent times the existence of veils or clouds over Mercury was not seriously challenged. In any case, nobody else had been able to carry out so long a study with so large a telescope. Consequently it was assumed that there must be an atmosphere of sorts, probably made up of the heavy gas carbon dioxide. This was borne out by reports from two more extremely eminent astronomers, Audouin Dollfus in France and V. Moroz in the U.S.S.R., both of whom claimed that they had found traces of a carbon dioxide atmosphere with a ground pressure of a few millibars (that is to say, less than 1/1000 the pressure of the Earth's air at sea-level).

This seemed consistent enough, but surprises were in store. It was suggested that the evidence for a Mercurian atmosphere had been due to a mis-interpretation of the data. Then Harlan Smith and his colleagues at the McDonald Observatory, in Texas, re-examined the whole problem, and found absolutely no trace of carbon dioxide or anything else. At present it is too early to say definitely that Mercury is without atmosphere; but the evidence is pointing that way.

What, then, about Antoniadi's 'clouds'? We are reasonably confident that even if a trace of atmosphere remains, it is not dense enough to hold dusty particles in suspension, and the idea of active vulcanism has very little to recommend it. With regret, it seems that we must dismiss the 'clouds' as being errors on the part of the observer, particularly as they have not been confirmed since. The results with the forthcoming probes may show that Antoniadi was right after all; but personally I doubt it.

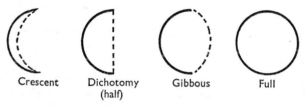

| Crescent | Dichotomy (half) | Gibbous | Full |

Fig. 17. Limb and terminator. The limb is shown as a solid line; the terminator is dashed.

It is best to admit, at once, that small or even moderate-sized telescopes are of no use for studying the surface features of Mercury. As the phase increases, so does the distance. We can never have a proper view of 'full Mercury', as the planet is then at superior conjunction and on the far side of the Sun. When at its brightest to the naked eye, Mercury is in the gibbous stage. The phases are not hard to see with a telescope (Fig. 17), and it is interesting to follow the changing position of the *terminator*.

The terminator is the boundary between the sunlit and the night hemispheres, and should not be confused with the limb, which is merely the edge of the apparent disk as seen from Earth. The difference is shown in the diagram, in which the terminator is dotted and the limb shown as a continuous line. In the case of the Moon, the terminator always appears rough and broken, with mountain summits catching the sunlight while adjacent valleys are still bathed in shadow. Mercury is so far off that its terminator appears smooth; but if we could see it better, the surface might look as rugged as a lunar landscape. The horns of the crescent are known as *cusps*. As long ago as

1800 a great German observer, Johann Schröter, noted that the upper or southern* cusp is usually blunter than the northern—presumably because the southern part of the disk is comparatively dark in hue.

This brings us on to the story of how Mercury was mapped, which is fascinating because it has provided us with a really major surprise.

The first serious telescopic observations of Mercury were made towards the end of the eighteenth century. Sir William Herschel, discoverer of the planet Uranus and an observer of genius, recorded no surface markings even with his powerful telescopes. Schröter, who was an amateur astronomer by inclination and chief magistrate of Lilienthal by profession, thought differently. He drew maps, and believed, quite reasonably, that he had seen definite and permanent markings. Unfortunately Schröter, though an honest and painstaking observer, was not a good draughtsman, and some of his reported 'discoveries', such as that of a mountain eleven miles high, are rather hard to credit. It is unlikely that he could have succeeded where Herschel had failed.

Between 1881 and 1889 Mercury was under scrutiny from Giovanni Schiaparelli, of Milan. Schiaparelli was noted for his keen sight, and he was using a good refractor. Rather than wait for sunset, when the planet was bound to be low in the sky, he preferred to make his observations in broad daylight; Mercury was then high up, though the presence of the Sun made the background inconveniently bright. Schiaparelli's map, given here (Fig. 18), was the first 'modern-type' attempt at charting Mercury. It remained the only one until Antoniadi's work over thirty years later.

Since then there have been various other maps, notably that of Camichel and Dollfus. Comparisons between them are interesting. There are various well-marked features, still known by the names which Antoniadi gave them; for instance, we have the Solitudo Hermæ Trismegisti (Wilderness of Hermes

---

* Since most astronomical telescopes give an inverted image, it has always been customary to orientate drawings and photographs with south at the top and north at the bottom. I have followed this time-honoured practice here, but recently the U.S. space authorities have begun to publish their charts with north at the top. So far this refers only to the Moon and Mars, but it has to be borne in mind when looking at the publications issued from all American astronautical offices; and the custom is spreading fast!

Fig. 18. Schiaparelli's map of Mercury.

the Thrice Greatest), Solitudo Criophori and so on. There are greyish regions and brighter areas, which—so far as we can tell—have definite and permanent outlines.

Without pretending that the maps agree really well, the overall impression is fairly clear; and from Earth, it seems that the surface features always occupy the same positions on the disk. Schiaparelli noted this, and came to a startling conclusion. He suggested that Mercury must have a 'captured' or synchronous rotation period, in which case it would spin once on its axis in the same time that it takes to go round the Sun: 88 Earth-days. The effects on the Mercurian calendar would be decidedly odd.

Consider the case of the Moon, which has an orbital period of 27·3 days. Its axial rotation period is exactly the same. Neglecting some minor effects, this means that the Moon keeps the same face turned Earthward all the time.* There is no mystery about it; tidal friction over the ages has been responsible. Originally, no doubt, the Moon rotated much more quickly, but the Earth's tidal pull slowed it down, until relative to the Earth (though not relative to the Sun) the rotation had ceased. If Mercury behaved in the same way with respect to the Sun, as Schiaparelli believed, there would be everlasting day over one hemisphere of the planet and permanent night over the other. Nothing of the kind had occurred to earlier observers. Schröter, for instance, had taken the rotation period to be about the same as that of the Earth.

Antoniadi was in full agreement with Schiaparelli, and the 88-day rotation period became so well established that nobody felt inclined to question it. It seemed, moreover, to be the only one which would fit the maps. But then, in 1962, came some disquieting news. At Michigan, in the United States, W. E. Howard and his colleagues measured the long-wavelength radiations coming from Mercury, and found that the dark side was much warmer than it could possibly be if it were always turned away from the Sun. Something was badly wrong.

At this point, radar astronomy was called in. Radar consists, basically, of sending out a radio pulse, 'bouncing' it off a solid body (or something else which acts in an equivalent way), and recording the 'echo'. Mercury is a small, elusive target, but by the mid-1960s it was well within radar range. When pulses are reflected from a body which is in rotation, the 'echo' is affected, and the rate of spin can be found. This was done for Mercury, initially by Rolf Dyce and Gordon Pettengill, using powerful equipment set up at Arecibo in Puerto Rico. They found that the rotation period must be far less than 88 days, and the modern value, now generally accepted, is 58·7 Earth-days.

Why, then, did the visual observers such as Schiaparelli and Antoniadi fail to record details on Mercury's 'other side'? To reject their work *in toto* seemed to be much too drastic, and a

* If you want a simple demonstration, put a chair in the middle of the room and then walk round it, turning so as to keep your face turned to the chair all the time. When you have completed one circuit, you will have turned on your axis (because you will have faced every wall of the room), but anyone sitting on the chair will never have seen the back of your neck. Similarly, from the Earth we never see the 'back' of the Moon, and until the first circum-lunar probe, Russia's Lunik 3 of 1959, we had no direct information about it. Note, however, that day and night conditions are the same for both hemispheres of the Moon.

solution had to be found. Now, at last, we believe that we have the answer. It lies in a curious relationship which may or may not be due to sheer coincidence.

I do not want to delve into mathematics, even very simple ones, so I will sum up matters as concisely as possible and then try to give the overall results. The facts are as follows:

1. The synodic period of Mercury—that is to say, the time which elapses between successive appearances at the same phase—is, on average, 116 Earth-days. If Mercury is 'new' on a particular date, it will again be 'new' 116 days later, and so on.

2. The rotation period (58·7 Earth-days) is equal to two-thirds of the revolution period (88 Earth-days).

3. It follows that to an observer at a fixed position on Mercury, the interval between one sunrise and the next would be 176 Earth-days, or 2 Mercurian years.

4. This interval, 176 Earth-days, is approximately equal to 1½ synodic periods.

5. From this, it will be found that after every 3 synodic periods, the same face of Mercury will be seen at the same phase.

6. Now for the coincidence—if coincidence it be! Three synodic periods of Mercury add up to approximately one Earth year. Consequently, the most favourable times for looking at Mercury recur every three synodic periods. Glance back at Point 5. You will realize that every time Mercury is best placed for observation, we see the same hemisphere, with the same markings in the same positions on the disk.

Antoniadi, of course, made the best use of his opportunities, but even with the great Meudon telescope the surface features on Mercury are hard to make out. He could study them best every third synodic period—and each time, he saw the same markings: in fact, those drawn on his map. He could not be expected to look for so curious a relationship between Mercury's behaviour and our own, so that it was natural for him to regard the rotation as being of the synchronous variety. At least it shows that there is no need to discard either his map or that of Schiaparelli. There was nothing basically wrong with the observations; it was the interpretation which was at fault.

Anyone who consults a set of tables, or who does some quick calculations, will realize that the coincidence is not exact. Yet it was good enough to deceive all observers up to the last decade, and many astronomers feel that the Earth's gravitational effects have a good deal to do with it.

We are now in a position to work out the weird calendar of Mercury, but things are complicated by the fact that we are still uncertain about the tilt of the axis. The Earth's axis is tilted to the perpendicular to the orbital plane by $23\frac{1}{2}$ degrees. It is usually believed that the axial inclination of Mercury is much less, and I propose to assume this in the description which follows, though it may turn out to be wrong.

Mercury's path round the Sun is decidedly eccentric, and the distance ranges between $28\frac{1}{2}$ million miles at perihelion to $43\frac{1}{2}$ million miles at aphelion. In accord with the traffic laws of the Solar System, the planet moves at its fastest when near perihelion; it can then attain 36 miles per second, though at aphelion this drops to a mere 24 miles per second. Near perihelion, the orbital angular velocity exceeds the constant spin angular velocity, so that an observer on Mercury would see the Sun slowly retrograde or 'move backwards' through rather less than its own apparent diameter for eight Earth-days around each perihelion passage. The Sun would then almost hover over what may be called a 'hot pole'. There are two hot poles, one or the other of which will always receive the full blast of solar radiation when Mercury reaches perihelion; the intensity must be two and a half times that absorbed by regions of the surface 90° of longitude away. Bear in mind, too, that from Earth the Sun has an apparent diameter of only half a degree; from Mercury, it will range between 1·1 and 1·6 degrees (Fig. 19).

Let us consider two observers, both of whom are placed on

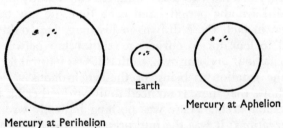

Earth

Mercury at Aphelion

Mercury at Perihelion

Fig. 19. Relative apparent size of the Sun as seen from Mercury at Perihelion and at Aphelion and from the Earth.

Mercury's equator, but who are 90 degrees in longitude away from each other. Observer A is at a hot pole, so that the Sun is at his zenith, or overhead point, at perihelion. This means that the Sun will rise when Mercury is near aphelion, and the solar disk will be at its smallest. As the Sun nears the zenith, it will slow down and grow in size. It will pass the zenith, and then stop and move backwards for eight Earth-days before resuming its original direction of movement. As it drops toward the horizon it will shrink, finally setting 88 Earth-days after having risen.

Observer B, 90 degrees away, will see the Sun at its largest near the time of rising, which is also Mercury's perihelion. Sunrise itself will be curiously erratic, because the Sun will come into view and then sink again until it has almost vanished. Then it will climb into the sky, shrinking as it nears the zenith. There will be no 'hovering' as it passes overhead, but sunset will be protracted; Mercury is back at perihelion, so that the Sun will disappear, rise again briefly as though bidding adieu, and then finally depart, not to rise again for another 88 Earth-days.

Another odd fact is that the stars will move across the sky at roughly three times the average rate of the Sun. And if the axis is more sharply tilted than most astronomers think, the polar conditions will be weird in the extreme, with several sunrises and sunsets per day!

One hates to think what a Mercurian would make of all this. Regrettably, however, we can discount the idea of any life on the planet. Conditions are hopelessly unsuitable in every way. Quite apart from the lack of atmosphere, there is the climate to be borne in mind. Most places will have two 'heat maxima' per day, one when the Sun is overhead and the other at the time of perihelion. A thermometer would rise to around 700 degrees Fahrenheit, so that metals such as tin would melt. Neither is the surface protected from dangerous short-wave radiations from the Sun. Whether manned landings can ever be made there is open to doubt; certainly they are out of the question in the foreseeable future.

For the moment, we can do no more than speculate as to the nature of the surface features. It is logical to think that Mercury is the same sort of world as the Moon, in which case the greyish patches will resemble the waterless lunar 'seas', while the brighter areas will be uplands. My own prediction, which I make with some diffidence, is that when automatic probes pass by and send back television pictures we will find that there are craters—but less definite and less well-marked than those of the Moon, because of the effects of the intense daytime heat. Crater walls will tend to crumble, and in any case there may be fewer structures. Within the next few years we ought to know whether I am right or wrong.

Meantime, the incomplete evidence available to us shows that the surface layer also is likely to be of the same type as that of the Moon. And like the Moon, Mercury is a very poor reflector of sunlight; its albedo, or reflecting power, is a mere 6 per cent. The colour is usually said to be pinkish, though I cannot claim to have detected any colour at all either telescopically or with the naked eye, and I tend to be sceptical.

Transits, when Mercury passes directly between the Sun and the Earth, are not particularly common, simply because the orbit of Mercury is inclined to ours at an angle of 7 degrees. The last transit took place on May 9, 1970; the next will be on November 10, 1973, November 13, 1986, and November 6, 1993. (Transits can occur only in May or November.) Strange phenomena have been reported during past transits, notably bright specks on the disk, but all these seem to have been due to defects in instruments or in the observer's eye.

When in front of the Sun, Mercury is too small to be seen without optical aid, and a telescope is needed. To look straight at the Sun through any telescope is to invite blindness; the only sensible method is to use the telescope as a projector, and view the Sun's image on a screen held behind the eyepiece. A 3-inch refractor is quite large enough to show Mercury well, and it is worth comparing the planet with any sunspots which may be on view. Mercury will look truly black, whereas a sunspot does not.

All in all, Mercury is one of the most hostile of all the worlds in the Sun's family. One day it is conceivable that men will go there, and see the huge, glaring Sun shining down from the darkness of the sky; but they will not stay for long. Mercury is a dead planet, as alien as it could possibly be.

Chapter Six

VENUS

VENUS, THE SECOND PLANET in order of distance from the Sun, is as different from Mercury as it could possibly be, apart from its high surface temperature. With its diameter of 7,700 miles, it is almost the same size as the Earth; it is an excellent reflector of sunlight, and it is the closest natural body in the sky apart from the Moon and an occasional passing asteroid. It moves round the Sun in an almost circular orbit at a mean distance of 66,200,000 miles, so that at its nearest to us it is only about a hundred times as far away as the Moon. At its most brilliant, it may even cast shadows.

Like Mercury, it stays in the same region of sky as the Sun, but the angular distance between it and the Sun can attain 47 degrees. This means that Venus can remain visible for as much as $5\frac{1}{2}$ hours after sunset, or rise $5\frac{1}{2}$ hours before the Sun does so. It can then be seen against a dark background, and is truly magnificent. Small wonder that the ancients named it in honour of the Goddess of Beauty.

Unfortunately it is a telescopic disappointment (Plate I), because the true surface is permanently hidden by a dense, cloudy atmosphere. Hard, sharp markings, such as those of Mars or even Mercury, are conspicuous only by their absence. Moreover, when Venus is at its closest to us, at the time of inferior conjunction, it has its dark side turned Earthward, so that we cannot see the planet at all (except during the rare transits). Full phase naturally occurs when Venus is on the far side of the Sun, so that again it is to all intents and purposes out of view. Maximum brilliance takes place when about 30 per cent of the sunlit side is turned in our direction. During the crescent stage, there is considerable evidence that keen-sighted people can make out the phase; good binoculars will show it easily (Fig. 20).

The phases themselves have been known for a long time, since they were recorded by Galileo during the winter of 1609–10. They can be predicted, since the orbit of the planet is known very accurately—and yet theory and observation seldom agree, as was first pointed out by the energetic Johann Schröter in the late eighteenth century.* Schröter made careful measurements of the time of dichotomy, when Venus appears as an exact half-disk. The results were surprising, since when Venus was an evening star, and therefore waning, dichotomy was always early; during morning apparitions, when the phase

Fig. 20. The changing apparent size of Venus; the planet is nearest when new, farthest when full.

was increasing, dichotomy was always late. There is no chance of the planet being out of position, and Schröter believed that because the brightness of the surface falls off near the terminator, one tends to think that the phase is less than it actually is. Against this, the discrepancy seems to vary from one apparition to another. Sometimes it is only a day or so, while at other elongations it may amount to about four days. My own observations from 1934 to the present time indicate that the mean value is two days.

* When I wrote a monograph about Venus, in 1956, I called this phenomenon 'Schröter's effect'. The term has now passed into official use, so that I have at least the satisfaction of having made a contribution to scientific language!

It is possible that the thick atmosphere of Venus may be responsible in some way; alternatively, Schröter's original explanation may be the correct one. There are even some observers who believe that the effect is due to nothing more significant than a trick of the eye. Amateurs can do useful work here, as I have indicated in the Appendix.

The atmosphere itself was first described in 1761 by M. V. Lomonosov, the first of Russia's famous astronomers. The occasion was one of the rare transits of Venus across the face of the Sun. The appearance of the planet's limb led Lomonosov to infer, quite correctly, that there must be an atmosphere of considerable depth and density.

Transits of Venus are interesting to watch with the naked eye —or so I am told; I have never seen one, since the last occurred in 1882. Transits occur in pairs separated by eight years, after which there are no more for over a century. The transits of 1874 and 1882 will be followed by those of 2004 and 2012.

During the seventeenth century Edmond Halley, the second Astronomer Royal—best remembered nowadays in connection with the comet which bears his name—improved on an earlier suggestion by James Gregory that transits of Venus might be used to measure the distance of the Sun, which in those times was not known with real accuracy. For this, it was essential to time the exact moment when Venus passed on to the face of the Sun, and to make observations from widely scattered points on the surface of the Earth. As the whole method is now completely obsolete there is no point in describing it in detail, but in any case the results were spoiled by an effect known as the Black Drop (Fig. 21). When Venus passes on to the Sun, it seems to draw a strip of blackness after it; when this strip disappears, the transit has already begun. The effect is due to the thick atmosphere around the limb of Venus, and nothing can be done about it. The 1874 and 1882 transits were well observed, but the results were most disappointing, and since there are now much better ways of working out the Sun's distance the next pair of transits will not be

Fig. 21. The Black Drop, seen at every transit of Venus.

regarded as of much

61

importance.* Very occasionally, Venus passes in front of a star, and hides or occults it; when this happens the star seems to fade for a few seconds before vanishing, because just before immersion its light is coming to us by way of the planet's atmosphere. The effect was well seen on July 7, 1959, when Venus occulted Regulus, in the constellation of Leo (the Lion). I watched it with a 12-inch reflector, and am glad to have seen it, since it will be centuries before Venus again passes in front of a really bright star.

Look at Venus through a telescope—even a powerful one— and you will see little more than a bright disk. If you are lucky, you may be able to detect a few shadings, but the markings are always very vague, and their outlines are bound to be indefinite. Moreover, they shift and change relatively quickly, so that they cannot be surface features; they are cloud phenomena in the upper atmosphere of Venus, and on the whole they tell us remarkably little. Neither are photographs of much help, though pictures taken in ultra-violet light do show some streaky markings. It is hardly surprising that before 1962, when the first successful space-probe by-passed the planet, our ignorance of Venus as a world was practically complete.

Theories were not lacking—and neither, for that matter, were maps. As early as 1732 an Italian, F. Bianchini, went so far as to publish a chart showing what he believed to be continents, oceans, bays and straits. Yet the small-aperture, long-focus refractor which he used was quite inadequate, and from a scientific point of view his 'map' was valueless. Even Schröter fell into the same sort of trap; and in our own century Percival Lowell, the American astronomer who is best remembered for his ideas about the canals of Mars, drew a strange network of

---

* All sorts of stories are connected with transits of Venus. Consider, for instance, the French astronomer G. Legentil in 1761 and 1769. The first of these was expected to be well seen from India, and accordingly Legentil set sail for Pondicherry. Unluckily for him, the Seven Years' War was raging, and he did not arrive until after the transit was over. Rather than risk a second delay, he decided to remain where he was for the next eight years, and to observe the 1769 transit instead. Shortly before and shortly after the vital hours the sky was brilliantly clear; but the transit itself was completely hidden by clouds, and as it was rather too long for Legentil to wait until the next transit (that of 1874) he packed up and set sail for home. Twice he was shipwrecked, and eventually reached Paris to learn that he had been presumed dead, so that his heirs were preparing to divide his property! The 1769 transit was notable because Captain Cook's epic voyage was made in order for observations of it to be carried out. Unlike Legentil, Cook and his astronomical assistants were successful. As an afterthought, they discovered Australia. . . .

hard, sharp, linear features which he regarded as permanent. Results of this kind must, alas, be rejected out of hand, and the observer who sets out to draw charts of the surface of Venus is apt to deceive himself as well as others.

Before 1962, it was thought that the atmosphere of Venus must be of about the same depth and density as that of the Earth, though it was known to be made up chiefly of carbon dioxide rather than oxygen and nitrogen. There was a widespread belief that the 'clouds' were due to nothing more exotic than water vapour, and the rotation period was thought to be about a terrestrial month, in which case there would have been seven or eight 'days' in every Venus 'year'. (Venus takes 224·7 Earth-days to complete one journey round the Sun.) Some astronomers believed the surface to be bone-dry and fiercely hot; others preferred the idea of a planet covered largely with water, in which case primitive life-forms could not be ruled out. One interesting result of this would have been that the carbon dioxide in the atmosphere would certainly have penetrated the water, so producing seas of soda-water; but nobody really knew much, and Venus was appropriately nicknamed the 'planet of mystery'.

Space-planners were keenly interested. Though there seemed no chance of intelligent life on Venus, there was always the hope that conditions there might not be utterly hostile, and in some ways Venus was thought to be more promising than Mars. The only way to find out was to use rocket probes, and in 1962 the Americans had their first success. Mariner 2 was launched on August 17; on December 14 it passed within 21,000 miles of the planet, and sent back a tremendous amount of information, much of which was frankly unwelcome.

Temperature measurements showed that Venus is extremely hot. The surface heat is over 700 degrees Fahrenheit, which does not look very favourable for the development of advanced life-forms. The idea of wide oceans had to be given up, since at that temperature water would promptly turn into steam, despite the high atmospheric pressure. The rotation period was found to be very long, and there was no detectable magnetic field. I propose to deal with these two latter points in some detail, because both are of exceptional interest.

If a planet shows permanent surface markings, the axial

rotation period can be measured without the slightest difficulty, merely by watching them as they are carried across the disk. Mars is convenient in this respect, and so, for that matter, is Jupiter. No useful results had ever been obtained for Venus, because the cloudy shadings are too variable and too vague. Many attempts had been made, but none could be regarded with any confidence at all;* estimates had ranged between 24 hours and as much as 224·7 days. If this last period were correct, then Venus would have a captured or synchronous rotation, with everlasting sunshine over one hemisphere and no daylight at all over the other. This did not seem out of the question, particularly since in 1962 Mercury was believed to spin in this fashion. Yet Mariner 2 indicated that the true rotation period of Venus was of the order of 250 days, actually *longer* than one Venus year. It also seemed that the planet rotated in a retrograde or wrong-way direction: east to west instead of west to east.

The whole situation appeared so odd that many astronomers tended to be sceptical. However, it has been confirmed by all subsequent research, both by space-probes and by radar measurements carried out from the surface of the Earth. The value for the rotation period accepted today is 243 to 244 days, with an axis which is almost perpendicular to the plane of the orbit. We come at once to another relationship which is too exact to be coincidental, and which shows that the Earth has had a pronounced effect upon the spin of Venus. The planet rotates exactly four times between successive inferior conjunctions, so that at each inferior conjunction the same hemisphere of Venus is turned toward us, and the rotation is 'locked' with respect to the Earth.

Before leaving this particular problem, it is worth noting that the French astronomers at the Pic du Midi, in the Pyrenees, have published some visual and photographic observations which indicate a rotation period which is only four days. They have not met with much outside support, but their results merit the most serious consideration, and the last word on the whole subject may not yet have been said.

Another surprise was the failure of Mariner 2 to detect any

* L. Brenner, in 1896, had given a period accurate to within 1/1000 of a second— which is rather like giving the age of the Earth to the nearest minute. Shades of Archbishop Ussher!

magnetic effects round Venus. The Earth has a relatively strong field, and there had seemed no reason to doubt that Venus, too, acted as a powerful magnet. There was even a certain amount of direct evidence. The Sun sends out streams of electrified particles, some of which reach the Earth; the Dutch astronomer Houtgast had found that each time Venus passes more or less between the Earth and the Sun, at inferior conjunction, there are measurable effects upon the number of solar particles bombarding us. Presumably, then, the particles had been diverted by the magnetic field of Venus.

Quite apart from this, there was the celebrated Ashen Light, which still causes heated arguments. This Light is the faint visibility of the 'night' side of Venus when the planet is at the crescent stage. The same sort of aspect is obvious enough in the case of the Moon, but can be explained easily as being due to light reflected on the Moon from the Earth. (Leonardo da Vinci realized this, a very long time ago.) Venus, however, has no satellite; reflected earthlight would be hopelessly inadequate to cause any visible glow, and it was widely believed that the Ashen Light must be produced by auroræ in the upper atmosphere of Venus. I shall say more about auroræ when we come to discuss the Earth. Meantime, it is enough to note that they are due to particles sent out from the Sun. Since they carry electrical charges, they are attracted to the magnetic poles, and are best seen from high latitudes.

The existence of auroræ over Venus would not prove the presence of a magnetic field, but it would support the idea. Unfortunately Mariner 2 showed no magnetic effects at all, and neither have the later probes, so that at best the magnetism of Venus must be much weaker than ours.

The Ashen Light was first reported by a Jesuit astronomer, Riccioli, in 1643, and has been observed at some time or other by almost everyone who has looked at Venus often enough with an adequate telescope. It is always faint, and it is by no means always visible. It has been summarily dismissed as a mere contrast effect, but in my view it is a real phenomenon; I have seen it too often, and too clearly, to believe otherwise. The auroral theory may yet prove to be correct, magnetism or no magnetism. With regret, we must dismiss the hypothesis put forward during the last century by Franz von Paula Gruithuisen, who explained the Light as being caused by vast bonfires lit by the local inhabitants to celebrate the election of a new Government!*

Mariner 2 was the pathfinder, but it did not provide all the answers, and some of its findings were regarded with suspicion. The next steps were taken by the Russians. Up to 1965 they had had no success with their planetary probes—not because their launching techniques failed, but because they had been unable to keep in touch with vehicles a long way away. Their first probe, Venera 1, had actually been dispatched in 1961, but signals from it ceased after a few weeks, and nobody knows what happened to it.

Veneras 2 and 3 were launched from the U.S.S.R. in November 1965, and both reached the neighbourhood of Venus in the following February. No. 3 is thought to have crash-landed, but once again the communications system failed, and No. 2 was no more informative. It was not until October 18, 1967 that the Soviet team had its first real triumph. On that date Venera 4 made a controlled landing on the planet; the capsule, containing the instruments, was separated before entering the atmosphere, and came down by parachute, transmitting clearly as it did so. On the following day, October 19, the American probe Mariner 5 by-passed Venus at 2,500 miles, after which it continued in a never-ending elliptical orbit round the Sun.

When the results from the two vehicles were compared, it was found that they were not consistent. In some ways they agreed; for instance, both indicated that the atmosphere of Venus contains at least 90 per cent of carbon dioxide. On the other hand, the Russians gave a much lower surface temperature and pressure than the Americans. It took some time for the situation to be sorted out, but we now know what happened. The pressure in the lower part of Venus' atmosphere was great enough to crush the instrument compartment of the Soviet probe, so that when its signals ceased Venera 4 had not reached the surface at all; it was still at least eighteen miles up. Therefore, its data referred not to the actual ground, but to the lower

---

* This was only one of Gruithuisen's extravagant ideas. He also announced the discovery of an artificial structure on the Moon, which turned out to be nothing more significant than a few low ridges. He was a keen-eyed observer, but his vivid imagination tended to discredit his work even during his lifetime.

atmosphere. When the necessary adjustment was made, the results were in fairly good accord with those from Mariner 5.

In January 1969 Veneras 5 and 6 went on their way. Both made soft landings, on May 16 and 17 respectively, and both kept on transmitting until they were at least very close to the surface. We may now be sure that the surface temperature is over 800 degrees Fahrenheit, and that the atmospheric pressure at ground level is more than 80 times as great as that of our own air; it may be as high as 140. Venera 5 gave rather more moderate readings than its companion, and there are suggestions that it came down on a mountainous region rather than in a depression. The last probe in the series, so far, has been Venera 7, which landed on Venus on December 15, 1970, and transmitted from the actual surface for almost half an hour. It gave a temperature reading of 887 degrees Fahrenheit, which was not unexpected. All in all, the Venera programme has been an outstanding success.*

If the percentage of carbon dioxide in Venus' atmosphere is between 93 and 97, there is not much room left for anything else; a little nitrogen is present, and there may be a trace of oxygen, but the amounts are negligible. And whatever the clouds may be, they are not watery. Crystals of solid carbon dioxide may be a more logical answer.

Venus has been called the Earth's twin; but why are conditions there so unlike those on our world? An interesting theory has been put forward recently by S. F. Singer, in America, who suggests that in the remote past Venus was hit by either a large asteroid or a small planet. If so, then the result might have slowed down the rotation from a conventional 24 hours or so down to its present leisurely rate. And if the intruder plunged through the crust of Venus, so much hot material might have been forced out of the interior that the whole evolutionary sequence was changed. It all sounds somewhat drastic, but it does at least provide some sort of explanation for both the slow rotation and the high temperature.

* It is surprising that failures have been so rare, because to land a probe on another planet calls for amazing accuracy. If the launching velocity is wrong by only three feet per second, the vehicle will miss Venus by something like half a million miles—unless, of course, a mid-course correction is made. This is no place to start a discussion about rocket techniques, but the point is worth making, if only to show how rapid our progress has been during the past decade.

Alternatively, it may be that the surface is so hot merely because Venus is closer to the Sun than we are. The atmosphere of the Earth was probably produced by gases sent out from below the crust. The water vapour in this mixture of gases duly condensed, first as rain and then in the low-lying regions which became our oceans. With Venus, however, it is suggested that the temperature was just too high for the water vapour to condense out, so that it remained in the atmosphere, and oceans could never develop. The lack of liquid water allowed carbon dioxide to accumulate in the atmosphere as well—instead of being kept at low levels, where it would combine with the minerals in the planet's crust. It was a kind of vicious circle, and resulted in a further rise of temperature, so that Venus has never had a cool surface.

The weakness here is that there does not seem to be much water-vapour in Venus' atmosphere now. Yet our analyses are still incomplete, and there may be more moisture than we think. Note, too, that if this last theory is correct, the Earth only narrowly avoided the same fate. Had it been a few million miles closer to the Sun, it, too, would have been too hot for the water-vapour to condense after it had been sent out from the Earth's interior—and you and I would not be living here today.

What the various probes have not done, so far, is to send back any television pictures; in any case, photography of the surface would be virtually impossible. Luckily, radar methods have come to our aid, and in America a research team headed by R. M. Goldstone and H. Rumsey has produced a map of part of the surface which, they say, gives a resolution twice as good as that obtainable of the Moon with the naked eye. The chart is shown here (Fig. 22); although it is rough and incomplete, it is a great deal better than nothing. The chief feature is the roundish patch known provisionally as Alpha. It is about 600 miles across, and may be a mountainous area, though opinions differ. Various smaller features are also shown, and there are some circular regions which are thought to be relatively smooth. Comparable results have been obtained by D. B. Campbell and his colleagues, using the 1,000-foot non-steerable 'dish' at Arecibo in Puerto Rico. They too have recorded Alpha, together with areas of low radar reflectivity. One of these, close to Alpha, seems to have a central feature which

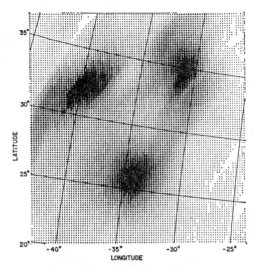

Fig. 22. Radar map of Venus.
(Goldstone.)

has been compared with the central peak of a lunar crater—
though it would be highly premature to infer from this that
Venus has craters, lunar-type or otherwise.

This is really as far as we can go at the moment, but there is
nothing against speculating about the conditions which will be
met by any courageous astronauts who land on Venus in the
future.

Day and night will be utterly unfamiliar (Fig. 23). The
length of the 'solar day' is 117 Earth-days; the Sun will rise in
the west and set in the east; and of course the Earth will have
an apparent motion of its own. Let us put ourselves in the
position of an observer on Venus who is looking upward at
the time of inferior conjunction. If he sees the Earth directly
overhead, he will know that after five sunrises and five sunsets
he will again find the Earth at the zenith at midnight.

Actually the Earth could never be visible from the surface of
Venus, because the thick, cloudy atmosphere will hide the sky
completely. There will be a strange, eerie half-light, and there
will be other peculiarities too. In a dense medium, rays of light
are bent, and at the surface of Venus they may be made to curve
more sharply than the curvature of the surface. It has even been

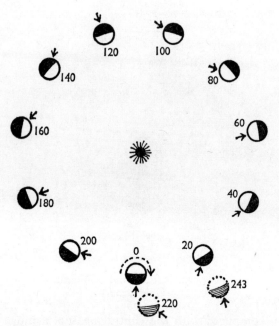

Fig. 23. The rotation of Venus. The rotation
period is 243 days, the sidereal period 224·7
days. The arrow indicates one fixed point on
the surface, which has a daylight period of
about 59 days followed by an equal period of
darkness.

suggested that an observer who looks straight ahead of him will
see the back of his own neck. This is an exaggeration, but the
general impression is bound to be weird, and in modern jargon
we may describe Venus as a psychedelic world.

Because the lower atmosphere is so dense, even a moderate
wind will have a great deal of force, and there must surely have
been tremendous erosion. Sharp mountain-peaks will have
been smoothed and levelled, if indeed they have ever existed.
There can be no similarity between the surface of Venus and
that of the Earth, or even Mars.

Half a century ago Svante Arrhenius, winner of a Nobel
Prize, was claiming that on Venus 'everything is dripping wet',
and that luxuriant vegetation flourished there, together with
amphibian and perhaps reptilian life. Even after the detection

of carbon dioxide in vast quantities, in 1933, it was still thought that some sort of life might survive. Nowadays we know better, and as a prospective colony Venus has proved to be a sad disappointment, so that in our list of targets it has yielded pride of place to Mars. There is no living thing on Venus, at least of the type we can understand. Instead of being pleasant and welcoming, the lovely 'Evening Star' has little to offer us.

Chapter Seven

THE EARTH

WHEN WRITING A BOOK about the Solar System, it is not easy to decide just how to deal with the Earth. It is a normal planet, and we tend to regard it as exceptional only because we happen to live here; but it is the concern of geophysicists rather than astronomers. In the present chapter, therefore, I propose to confine myself mainly to matters which have a purely astronomical slant.

There is nothing unusual about the Earth's orbit. Our average distance from the Sun is 92,957,000 miles; the sidereal period is 365¼ days, and the mean orbital velocity is 18½ miles per second, or 66,000 m.p.h. The Earth's path round the Sun is not perfectly circular; we reach perihelion in January, aphelion in July. The seasons are due not to the changing distance (91½ to 94½ million miles), but to the tilt of the Earth's axis, which amounts to 23½ degrees to the perpendicular to the orbital plane (see Fig. 24). In position 1, the northern hemisphere is tilted sunward, and Europe has its summer; in position 2, it is northern winter. The fact that the Earth is three million miles closer to the Sun in position 2 makes very little difference, and the effects on the world climate are more or less masked by the unequal distribution of land and sea in the two hemispheres.

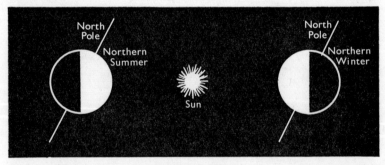

Fig. 24. The Seasons. In northern summer, the north pole is tilted toward the Sun.

Of the remaining planets, Mars, Saturn and Neptune have axial inclinations very similar to ours. Uranus has a much greater tilt, while that of Jupiter is much less; about Mercury, Venus and Pluto we have little or no positive information, though Venus seems to be an oddity in almost every respect.

In size and mass the Earth is equally unremarkable. It has a diameter of 7,926 miles as measured through the equator, but only 7,900 miles as measured through the poles, because the globe is appreciably flattened—rather less than that of Mars, but more so than with Mercury or Venus. The specific gravity is 5·5, so that the Earth 'weighs' 5·5 times as much as an equal volume of water would do. Mercury is of about the same density; Venus and Mars rather less, while the giant planets are so different in nature that comparisons are rather pointless.

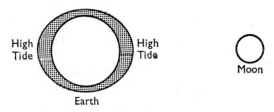

High Tide — High Tide

Earth

Moon

Fig. 25. The Tides. In this diagram, the depth of the water-shell is very much exaggerated in both extent and regularity.

In one respect the Earth really is unique. It is the only relatively small planet to be attended by a large satellite, and it is wiser to regard the Earth–Moon system as a double planet. Quite apart from its value as a source of illumination at night, the Moon is the main cause of the ocean tides so important to our shipping. The general theory is given in Fig. 25. As the Earth spins round, the Moon's gravitational pull tends to heap up the water in a bulge, with a corresponding bulge on the far side of the Earth. The water-heaps do not rotate with the Earth, but stay 'underneath' the Moon, so that they seem to pass right round the world once a day; since there are two heaps, each point on the Earth's surface has two daily high tides. In practice there are many complications, and the Sun also has a strong tide-raising effect; when the Sun and Moon are

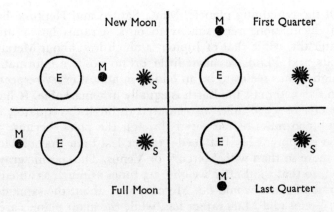

New Moon   M   First Quarter

Full Moon   Last Quarter

Fig. 26. *(Left)* Spring tides; the Sun and Moon are pulling in the same sense. *(Right)* Neap tides; the Sun and Moon are pulling at right-angles to each other.

pulling in the same direction (that is to say, at new or full moon) the tides are at their most forceful (Fig. 26).

Tides on other planets would be of different type—even assuming that there were any other planets with oceans upon them. Venus has no satellite, while any hypothetical Martian seas would be calm and sluggish, since Mars is much farther away from the Sun, and its two dwarf moons could be ignored as potential tide-raisers.

By now there are thousands of man-made bodies orbiting the Earth, and there have been unkind remarks about 'space junk'. Once an object has been put into a stable orbit which keeps it above the top of the resisting atmosphere, it will keep moving round us indefinitely. For instance Telstar, the satellite which provided the first global television relay, is still in orbit, though its power has long since failed and all trace of it has been lost. It was never visible with the naked eye, and we could never find it again even if there were any point in looking for it.

Yet can there be a second natural satellite? There is no theoretical reason why not; after all, some of the smaller satellites of other planets are quite unlike our massive Moon, and may well be captured asteroids.

The idea of a second moon is not new. Jules Verne used it in his famous space-travel story of a century ago; indeed, it was an essential part of his plot, since the wanderer pulled the man-carrying projectile out of its original path, and swung it round the Moon back to Earth. But even if a minor satellite exists, it must be very small. Assuming a reflecting power equal to that of the Earth (around 40 per cent), a 25-mile satellite as far away from us as the Moon would shine as brightly as most of the stars; it would be the equal of Betelgeux in Orion, so that it would have been known from the earliest times. Even at two million miles, a 25-mile body would be visible with the naked eye, while a 12-mile satellite would be conspicuous in binoculars. Even if we consider a satellite only one mile across, we find that a moderate telescope would show it out to several millions of miles, so that it would certainly have been detected long before now. In fact, any minor satellite must be extremely small—a mere lump of material, probably irregular in shape.

Shortly after the end of the war, Clyde Tombaugh, discoverer of the planet Pluto, carried out a long and systematic hunt for minor satellites. His equipment would have been capable of picking up a football-sized body a thousand miles away, even if the object had been made of dark rock, while a 10-ft.

diameter satellite would have been detectable out to 10,000 miles. Nothing whatsoever was found, which was a source of relief to the space-research authorities; a collision between a manned probe and even a small natural satellite would have disastrous results!

On the other hand, there is a suggestion that some loose 'clouds' of meteoritic material move round the Earth in the same path as the Moon, one cloud keeping well ahead of the Moon and the other well behind it

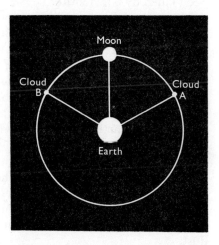

Fig. 27. 'Lagrangian points', 60 degrees away from the Moon but in the same orbit. At these points Kordylewski claims to have located two clouds of diffuse material.

(Fig. 27). Evidence produced by the Polish astronomer K. Kordylewski seems to be moderately convincing; but even if such clouds exist they must be of negligible mass, and personally I will be convinced of their reality only when somebody manages to photograph them.

Though we lack a second satellite, vast numbers of small particles enter the upper atmosphere each day, only to be destroyed by friction against the air-particles. These are the meteors, known more popularly as shooting-stars. The average meteor is no larger than a grain of sand, and can be seen only during the last second or two of its career, when it becomes hot enough to make its presence known. Meteorites —larger bodies, which survive the complete drop to the ground without being burned away—come into a different category; they are not simply large meteors, but are less friable, and seem to be more closely related to the asteroids. Oddly enough, the Earth's air is no real protection against them. One known meteorite, still lying where it landed at Hoba West in Africa, weighs over 60 tons, but major falls are fortunately rare, and there seems to be no authenticated case of anyone having been killed by a meteorite.

Meteorite falls are seldom seen, though admittedly there were many witnesses of the Barwell Meteorite of Christmas Eve 1965, which blazed across the English sky before breaking up and landing in Leicestershire; more recently there has been the Bovedy Meteorite, which also was widely observed during its last moments of flight.* On the other hand there can be few people who have not seen shooting-stars. Each August, for instance, the Earth plunges through a shoal of meteors, and the result is the spectacular shower known as the Perseids. Of course, no meteor can become luminous unless the air around it is sufficiently dense to cause heating by friction. Meteors begin to shine at about 120 miles above sea-level; higher up the air is too thin. The average meteor burns out while still forty or fifty miles up, and ends its journey in the form of fine dust.

Meteors would never appear over the airless Moon, and in all

* I missed it by about a minute. I had been in my observatory at Selsey, and had just gone indoors when the meteorite passed over. The main mass fell in the sea, but many fragments were found in Northern Ireland.

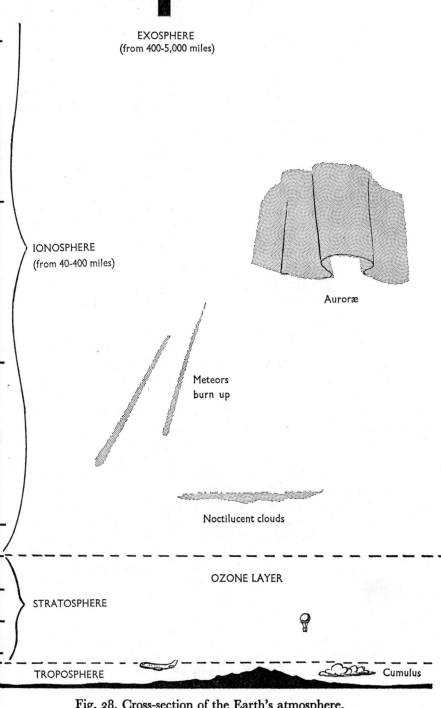

EXOSPHERE
(from 400-5,000 miles)

IONOSPHERE
(from 40-400 miles)

Auroræ

Meteors
burn up

Noctilucent clouds

OZONE LAYER

STRATOSPHERE

TROPOSPHERE   Cumulus

Fig. 28. Cross-section of the Earth's atmosphere.

probability they would be similarly absent from the sky of Mercury, while from cloud-covered Venus it would be impossible to carry out any astronomical observations at all. However, meteors would certainly be seen from Mars, where the atmosphere is quite dense enough to make them burn away.

Of the various glows which can be seen in our sky, I need say little here about the Zodiacal Light, which extends along the ecliptic, and is seen only for a short while after sunset or before sunrise, stretching upward from the horizon; it is due to sunlight reflected from thinly-spread matter in the main plane of the Solar System. The even more elusive Gegenschein, or Counterglow, may be due to a similar cause. But when we come to polar lights or auroræ, we are touching upon the realm of geophysics, because auroræ occur well inside the Earth's atmosphere.

The atmosphere itself is made up of several layers (Fig. 28). The lowest is the troposphere, in which we live; it extends up for about seven miles, and contains over three-quarters of the total atmospheric mass of about 5,000 million million tons. Above the troposphere comes the stratosphere, ranging between seven and forty miles; the temperature there is constant at about −67 degrees Fahrenheit, and there are high-velocity winds, though in these rarefied regions even a 100-knot 'gale' has little force. Continuing upward we come to the ionosphere, from forty to 400 miles; it is here that we find the layers which reflect some radio waves back to the ground, so making long-range wireless communication possible. Finally, above 400 miles, comes the exosphere. It has no definite upper limit, but simply tails off into space; traces of it may linger on out to 5,000 miles or even more.*

In comparing this sort of structure with those of other planetary atmospheres, we are more or less limited to Mars, because we still know relatively little about Venus. Though it is too early to be sure, it seems that Mars may have an ionosphere, in which case wireless contact between one 'Martian base' and another may be possible. On the Moon, and on Mercury, wireless range will be limited to the distance of the

---

* I have deliberately simplified matters, and ignored various often-used terms such as 'mesosphere' and 'thermosphere'. All I have tried to do is to give a general picture.

horizon, and for communications purposes on the Moon it will be desirable to send up orbiting stations of the Telstar variety.

Next, let us turn to the magnetosphere, which may be defined as the region of the Earth's magnetic field. It is rather like a tear-drop in shape, with the tail pointing away from the Sun. On the sunward side of the Earth it extends for only about 40,000 miles, but on the dark side it reaches out much further. The Sun sends out streams of particles, making up what we call the solar wind. When these particles come toward the Earth they meet the magnetic field, and produce a shock-wave.

Inside the magnetosphere there are two zones of intense radiation, known as the Van Allen zones in honour of James van Allen, the American scientist who was principally responsible for their discovery. They were detected with instruments carried aboard Explorer 1, the first successful U.S. satellite, which went up on February 1, 1958. There are two main belts, one with its greatest intensity at 3,000 miles and the other at 10,000 miles; the inner is made up chiefly of protons, while the outer consists mainly of electrons. Probably they make up one system rather than two completely separate ones.

This may seem a digression, but it is not really so, because it is bound up with the cause of auroræ. When the Sun is active, very energetic particles sent out from it arrive in the Van Allen region. What follows is not entirely clear, but it may well be that the Van Allen zones become 'overloaded', and particles cascade down into the lower exosphere and even the ionosphere, causing the lovely glows. Since the particles carry electric charges, they make for the Earth's magnetic poles, which explains why auroræ are best seen from high latitudes. A night in North Norway, Hudson's Bay or Antarctica would seem drab without them, but they are much less frequent from South England or New York, and from the Mediterranean and all places closer to the equator they are rare enough to attract general attention whenever they are seen.

No auroræ would shine over Mercury or the Moon. From Mars they might occur, but would probably be less vivid and colourful than ours. As we have noted, there are suggestions that the Ashen Light of Venus may be due to intense auroral activity there; but nobody on the surface of that highly uninviting planet would have any hope of seeing them. Going further afield, there is excellent evidence that the magnetic field of Jupiter is very powerful, and the same may apply to the other giants. Certainly the Earth is not unique in possessing a magnetosphere, though as yet no fields have been detected round Venus or Mars, and the lunar magnetic field is so weak that we are entitled to ignore it.

In one way the Earth's atmosphere is awkward scientifically, as it blocks out most of the radiations coming towards us from outer space. Only visible light, together with some shorter wavelengths and a certain amount of long-wavelength radio emission, can pass through—which is one reason why astronomers are so anxious to take their equipment (and, ultimately, themselves) above the screening layers of the ionosphere. From this point of view conditions will be better on the Moon, where there is no barrier at all. Yet, but for the layers in the ionosphere, no life here could ever have developed, because some of the short-wave radiations are lethal in large doses. This may be one reason for regarding Mars as sterile; the atmosphere there seems to be too tenuous to provide an effective screen.

When we turn our attention downward instead of up, we have to revert to theory. It has been said that we know less about the core of the Earth than we do about the interior of the Sun, though this is no longer strictly true. The diagram shows the modern view of the way in which the world is made up (Fig. 29). There is a central core, made of nickel-iron; it is usually thought to be liquid, though the tremendous pressure may turn the innermost part of it into a solid. Outside the 3,600-mile core comes the mantle, composed

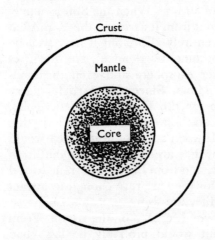

Fig. 29. Structure of the Earth. The depth of the crust is indicated by the breadth of the solid line!

chiefly of the volcanic rock peridotite. And outside the mantle comes the crust—no thicker, relatively, than the skin to an apple.

Much of our knowledge about the Earth's interior has been gained from a study of the waves set up by earthquake shocks. This is no place to discuss them in any detail, but basically there are two kinds of earthquake waves which concern us here. One type can travel through a liquid, while the other cannot. Our measurements of the size of the liquid core have been made by finding out just where waves of the second type are stopped. We would very much like to carry out similar experiments on the surface of the Moon, and a start has been made inasmuch as results from the earthquake-recorder left there by the astronauts of Apollo 12 have been most illuminating.

Not, of course, that the Moon has a terrestrial-type iron core; probably it lacks anything of the kind, since its overall density is about the same as that of the Earth's outer layers. At the core of the Earth, the specific gravity must be well over 10, with a temperature exceeding 4,000 degrees Fahrenheit. Whether the Moon and Mars are hot or cool at their cores is still not known; Venus must presumably be hot, since in size and mass it is strikingly like the Earth, and the differences between the two planets may be more superficial than fundamental.

Geology can tell us a great deal about the evolution of the Earth. It is the very beginning of the story which remains somewhat obscure. Probably the Earth lost all its original atmosphere, and after a while a new atmosphere was produced by gases sent out from below the crust. There must have been intense volcanic activity, on a scale which we find difficult to visualize. At first the atmosphere was composed largely of water-vapour and carbon dioxide, but then the bulk of the water-vapour condensed out, leading to the formation of the oceans. With the spread of plant life on to the continents, the whole situation changed, since the plants extracted much of the carbon dioxide from the atmosphere and replaced it with oxygen by the familiar process known as photosynthesis. If we could go back to, say, the Cambrian Period, 500 million years ago, we would find ourselves unable to breathe. Land plants had not appeared, so that there would be too much carbon dioxide in the atmosphere.

The sequence of events on Venus has been obviously different, and Singer's rather drastic theory, described earlier, is attractive even though it is so speculative and may turn out to be very wide of the mark. Mars, with a much lower mass and escape velocity, may have produced a secondary atmosphere in the same way and at about the same time as the Earth, but was unable to hold it down. And with the giant planets, the gravitational pulls are so strong that all the original hydrogen-rich atmosphere has been retained. At any rate, the Earth is the only planet in the Solar System to have an atmosphere suitable for our type of life.

From the Moon, the Earth is a glorious sight, as the Apollo astronauts have told us (Plate VI). From Mars it would be an inferior planet, showing lunar-type phases, and moving much as Venus seems to do to us. An observer as far out as Jupiter would have difficulty in seeing the Earth at all, and from the more remote planets it would be lost to view in the Sun's glare. In size and mass it is of minor importance in the Solar System; but it is our home, and it suits us well.

Chapter Eight

THE MOON

THE MOON HAS BEEN reached; men have walked there, watched by television viewers all over the world. I venture to predict that the year 1969 will be remembered long after 1066 has been forgotten. The landing of Neil Armstrong and Edwin Aldrin proved that the age of terrestrial isolationism is at an end.

I have heard it said that the Apollo journeys have robbed the Moon of some of its romance. This may be true—but from a scientific point of view the Moon is more intriguing than anyone could have believed. Admittedly there is no life there, but the whole lunar world is full of surprises.

First, let us elaborate slightly on the Moon's status in the Solar System. To us it appears the most splendid object in the sky, apart from the Sun, and it is not surprising that ancient peoples worshipped it as a god. And yet the Moon is small by planetary standards. Its diameter is a mere 2,160 miles, and it owes its apparent eminence to the fact that it is so close to us. At its mean distance of only 239,000 miles, it is one hundred times as near as Venus.

I have said that in my view, the Earth–Moon system should

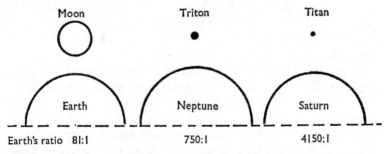

Fig. 30. Relative sizes of satellites compared with their primaries. The primaries are drawn to different scales. Thus Titan is much larger than our Moon—but it is much smaller relative to Saturn, which is so much larger than the Earth!

be regarded as a double planet rather than as a planet and a satellite. The Earth has 81 times the mass of the Moon, but with all other planet–satellite systems the discrepancy is much greater (Fig. 30). For instance, Titan, the senior member of Saturn's family, has only 1/4150 the mass of Saturn, even though it is larger than our Moon; remember that Saturn is a giant. With Neptune, the mass of the larger satellite (Triton) is only 1/750 of that of Neptune itself. Therefore our Moon is exceptional, and we are entitled to place it in a class of its own.

Until quite recently there was strong support for the idea that the Moon broke away from the Earth many millions of years ago. This theory was worked out in detail by G. H. Darwin (second son of Charles Darwin), and looked very plausible. According to Darwin, the Earth and the Moon used to be one body; quick rotation led to instability, and a large portion was thrown off, producing the Moon and leaving a hollow in the Earth's crust which was later filled by the waters of the Pacific Ocean.

Unfortunately for this theory, it has now been found that from a mathematical viewpoint the process will not work. A piece of material the size of the Moon could not be thrown off as Darwin had believed, and in any case it could never remain as a separate body; it would break up, and most or all of it would fall back to Earth.

In 1969 the theory was revived, albeit in modified form. This time it was suggested that Mars, too, could have been involved. If both Mars and the Moon were thrown off the Earth, there are fewer mathematical objections: the Moon would be the 'droplet' left behind by the departing mass which became Mars. But the idea is highly speculative, and has met with little support.

If we agree that the Moon and the Earth have always been separate, then there are only two possibilities. Either the Moon used to be an independent planet, and was captured when it happened to wander close to the Earth, or else it and the Earth were born at about the same time and in the same region of space, so that they have always remained together. The problem is difficult to solve, though when we learn more about the Moon's inner composition we may be in a better position to tackle it.

It is often said baldly that 'the Moon revolves round the Earth'. In a way this is true enough, but it does not tell us the whole story. Strictly speaking, the Earth and the Moon revolve round their common centre of gravity, much as the two bells of a dumb-bell will do when twisted by their joining arm. Since the Earth is 81 times as massive as the Moon, this centre of gravity or *barycentre* is displaced toward the Earth; in fact it actually lies inside the terrestrial globe, so that our simple original statement is good enough for most purposes so long as we remember that the mass of the Moon is far from negligible.

Everyone is familiar with the phases of the Moon. They must have been known from the earliest days of human history, and there is no mystery about them once we realize that the Moon shines only by reflected sunlight. The diagram in Fig. 31 is self-explanatory; the Moon is new in position 1, half at 2, full at 3, and half again at 4. Between 1 and 2, and again between 4 and 1, it is a crescent; between 2 and 3, and between 3 and 4, it is gibbous. The only points worth adding are that the true 'new moon' is invisible, and that positions 2 and 4 are known as First Quarter and Last Quarter respectively.

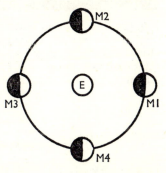

Fig. 31. Phases of the Moon. New: M1. First Quarter: M2. Full: M3. Last Quarter: M4. The sunlight is assumed to be coming from the right.

The sidereal month, or time taken for the Moon to complete one journey round the Earth (or, more accurately, round the barycentre) is 27·3 days. However, the interval between successive new moons is rather longer, because the Earth is itself moving round the Sun. In the next diagram (Fig. 32), suppose that the Moon is represented by M, the Earth by E, and the Sun by S. In the upper position the Moon is new at M1, since it lies between the Earth and the Sun. After 27·3 days the Moon has arrived back at M1, but it is not now lined up with the Sun, because the Earth has moved along its orbit. Only when the Moon has arrived at position M2 will it again be 'new'. The *synodic month,* or interval between

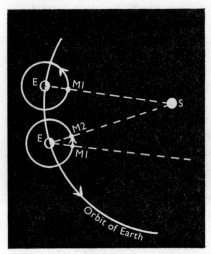

Fig. 32. The interval between successive new moons is longer than the Moon's sidereal period of 27·3 days. In the upper diagram the Moon is new at M1; but in the lower diagram it is not new until it has passed M1 (after one full circuit of the Earth) and has reached M2, because in the interval the Earth has moved on round the Sun.

successive new (or full) moons, is therefore 29·5 days instead of only 27·3.

At full phase, the Moon sometimes passes into the shadow cast by the Earth. The result is a lunar eclipse; all direct sunlight is cut off from the Moon's surface, and the Moon becomes very dim until it emerges from the cone of shadow. It does not disappear completely, because some of the solar rays are bent on to the lunar surface after having passed through the atmosphere of the Earth.

Lunar eclipses (see Fig. 33) may be either total or partial, and are more often seen than eclipses of the Sun.* They may be spectacular, and are always worth looking at. Their scientific importance lies in the fact that as the Earth's shadow sweeps over the Moon, and causes a wave of sudden cold there, certain areas show a fall in temperature which is much less than the average. One such region is the brilliant crater Tycho (Plate IV), about which I have more to say below. These areas are often known as 'hot spots', but the term is misleading inasmuch as during an eclipse, or during the lunar night, they are still very cold indeed.

The fact that the Moon's axial rotation period is also 27·3 days, so that it keeps the same hemisphere turned permanently Earthward, was intensely irritating to astronomers of

* Solar and lunar eclipses are about equally numerous; but a solar eclipse is visible only from a narrow strip across the Earth's surface, whereas a lunar eclipse, when it occurs, can be seen from an entire hemisphere. An observer who stays at any particular place on the Earth will therefore see more lunar eclipses than solar ones.

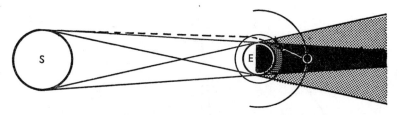

Fig. 33. Theory of a lunar eclipse. The Moon passes into the cone of shadow (black) during a total eclipse, but some light is refracted on to it through the Earth's atmosphere, as shown by the dashed line. The penumbra (shaded area) lies to either side of the main cone or umbra.

the pre-Space Age. Actually, rather more than half the entire surface could be examined. The Moon rotates at a constant angular velocity, but its orbital speed varies, because its path is not circular; it moves quickest when closest to the Earth (perigee) and slowest when it is farthest away (apogee) (Fig. 34). Therefore, the position in orbit and the amount of axial spin become periodically out of step. We can see alternately for some distance beyond first one mean limb and then the other, so that the Moon seems to rock very slowly to and fro; this 'libration in longitude' is evident from one night to the next. There is also a 'libration in latitude', because the Moon's orbit is inclined to ours at an angle of five degrees, and we can see for some distance beyond alternate poles. Finally there is 'diurnal libration', due to our being placed on the surface of the Earth, almost 4,000 miles from the centre of the globe; at moonrise we can peer a little way beyond the mean limb. The overall result of these librations is that from Earth we can see a grand total of 59 per cent of the Moon's surface, though, of course, never more than 50 per cent at any one time.

The remaining 41 per cent is permanently out of view, and in the past there were some strange suggestions about it. Hansen, a last-century Danish astronomer, even believed that the Moon might be lop-sided in mass, and that all the air and water had been drawn round to the far side, which might be inhabited! He met with practically no support from his colleagues, but right up to 1959 it was still thought possible that the hidden regions might be different in nature from those which we can see with our telescopes.

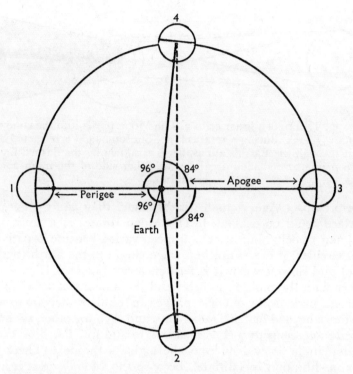

Fig. 34. Libration in longitude. Because the Moon's orbit is not circular, its orbital velocity varies; it moves from 1 to 2, and from 4 to 1, more quickly than from 2 to 3 or 3 to 4, whereas its rotational velocity remains constant. Therefore, the apparent centre of the visible disk as seen from Earth is displaced.

Then, in October 1959, the Russians dispatched Lunik 3, which went right round the Moon and sent back pictures of the far side. Today, these photographs seem very blurred and of poor quality; at the time, they represented a tremendous technical feat. Since then, orbiting lunar probes, both manned and unmanned, have provided us with photographs of the entire Moon, taken from close range. The only essential difference between the two hemispheres is that on the far side there are none of the broad, darkish plains which we still mis-call 'seas'.

The first telescopic maps of the Moon were produced in 1609–10. Priority must go to Thomas Harriot, sometime tutor to Sir Walter Raleigh, but the most detailed study was made by Galileo, whose map showed various features in recognizable form. It had long been suggested that the Moon must contain mountains and valleys; Galileo and Harriot saw them clearly, together with the walled circular structures which we call craters.

Any pair of good modern binoculars will show the details well, and with even a small telescope the sight is magnificent. First there are the 'seas' or *maria*, which are the smoothest parts of the surface even though by ordinary standards they are still extremely rough (as the astronauts know only too well). They have been given romantic names, such as the Mare Tranquillitatis (Sea of Tranquillity), Mare Nubium (Sea of Clouds), Mare Crisium (Sea of Crises), Oceanus Procellarum (Ocean of Storms), Sinus Iridum (Bay of Rainbows), Palus Nebularum (Marsh of Mists) and so on. Originally it was thought that they were true seas, or at least sea-beds. Studies of the samples brought home from the Moon have now shown otherwise; it seems that there has never been any liquid water on the Moon. Certainly there is no water now, and suggestions that there may be sub-crustal deposits of ice are not very convincing.

The Moon's lack of atmosphere and water make it a hostile place, but nothing else was to be expected. The lunar escape velocity is only $1\frac{1}{2}$ miles per second, so that an atmosphere made up of the same gases as our own would leak away quite quickly. Until recently it was thought that a trace might remain, but the astronauts have found no sign of it, and we must now concede that the Moon can be aptly described as an airless world. Without atmosphere, there can be no water. The lunar seas are dry lava-plains without a hint of moisture in them.

Many of the larger maria are basically circular in outline. Such is the vast, well-marked Mare Imbrium (Sea of Showers), which is easily visible with the naked eye, and has an area equal to that of Great Britain and France combined (Fig. 35). The walls of the circular maria are raised into mountain chains such as the Apennines, the Alps and the Caucasus. The Apennines, much the most spectacular of the ranges—though not the highest—form part of the border of the Mare Imbrium, and include peaks which attain 15,000 feet above the land below. There are many other ranges of the same kind, though some of

Fig. 35. Section of the author's 2-foot map of the Moon, showing the Mare Imbrium. The Apennines and Alps make up parts of the boundaries of the Mare. The dark-floored 60-mile crater Plato lies to the north (bottom) of the Mare. The Russian 'crawler' Lunokhod I has been operating in the area south of the Sinus Iridum or Bay of Rainbows.

them are less continuous. Isolated elevations abound; no part of the Moon is free from them.

The entire lunar scene is dominated by the craters, which range from huge enclosures well over 100 miles in diameter down to tiny pits too small to be seen from Earth. Copernicus, a 56-mile crater near the edge of the Mare Nubium, has massive, terraced ramparts rising to over 12,000 feet above a sunken interior upon which there is a group of central mountains. Not far away is Stadius, of comparable size, but low-walled and hard to trace; it gives every indication of having been overwhelmed by lava (though opinions about this differ). The structures are of many kinds. There are craters with central peaks, and others with relatively flat floors; craters with bright interiors, others with floors so dark that under suitable conditions of solar lighting they look like pools of ink; craters which are regular and well-defined, others which have been so broken and deformed that they are scarcely recognizable. Moreover, the craters frequently overlap each other, and they are spread all over the Moon, from the grey seas to the bright uplands and even the slopes and summits of mountain peaks.

In general, the craters are named after scientists of the past—a convenient albeit rather controversial system introduced in 1651 by a Jesuit priest-astronomer, Riccioli, who drew a lunar map which was reasonably accurate by the standards of the time. Astronomers such as Ptolemy, Tycho, Kepler and Copernicus are all there (though Riccioli, who had no use for the absurd idea that the Earth moves round the Sun, did admit that he had 'flung Copernicus into the Ocean of Storms'). Julius Cæsar is also among those present, though he has been honoured for his calendar reform rather than for his military prowess. Among the more unexpected names are Birmingham, Billy and Hell, all of which commemorate past astronomers of considerable eminence. Needless to say, Riccioli's system has been extended since 1651 to accommodate later scientists, such as Newton, Halley and Einstein. In 1970 the International Astronomical Union, the controlling body of world astronomy, ratified names which had been suggested for the craters on the newly-mapped far side of the Moon.

Though some of the craters are deep, they are not in the least like steep-sided mine-shafts, as is shown by the typical cross-section in Fig. 36. A normal crater has a rampart which rises to a modest height above the outer country, and a floor which is

Fig. 36. Cross-section of a typical large lunar crater.

depressed below the mean level of the surface. Where there is a central peak, we find that the summit never attains a height equal to that of the surrounding walls. Also, lunar slopes are surprisingly gentle. When astronauts land inside large craters, they will not feel in the least shut in, and anyone in the middle of—say—the 92-mile Ptolemæus, shown in the photograph, would be unable to see the walls at all, since they would be below his horizon. Because the Moon is so much smaller than the Earth its surface curves more sharply, and to an observer on the ground the horizon is only a mile or so away.*

The depths of the craters—and, for that matter, the heights of the mountains—are measured by the lengths of the shadows they cast. When the Sun is low over a crater, the interior will be filled with shadow, making the crater itself very conspicuous. Usually, a crater is at its most spectacular when almost on the terminator. The effect is perhaps best seen with the Sinus Iridum or Bay of Rainbows, which leads out of the Mare Imbrium, and is shown in Plate III. If it is observed at the right moment, the mountainous rampart toward the Moon's limb will seem to project out of the darkness, producing a scene which has been nicknamed 'the jewelled handle'.

It is interesting to follow the progress of sunrise over various parts of the Moon. A crater which is very prominent one night, when it is shadow-filled, may become very obscure later, when the shadows inside it have almost disappeared. At full moon, when the shadows are at their shortest, even large, deep craters such as Gassendi (Plate V) become hard to identify. Full moon is the very worst time to start trying to find one's way about, and the most spectacular views are to be had between the crescent and three-quarter phase.

Of course, there are some special features which are easy to recognize at any time. Such is Plato, a 60-mile crater with a dark floor which Hevelius, a mid-seventeenth-century lunar observer, called 'the Greater Black Lake'. Near the Moon's limb there are two more very dark-floored formations, Grimaldi and Riccioli.† All these are basically circular, but since they lie

---

* Quite apart from this, distance-judging is not easy on the Moon. The astronauts have found that remote objects tend to seem much closer than they actually are.

† No prizes are offered for guessing who named this particular crater, which is rather larger than Wales. Its even bigger neighbour is named in honour of Grimaldi, who was Riccioli's pupil.

well away from the apparent centre of the Moon they are fore-shortened into ellipses. Very close to the limb, foreshortening becomes so marked that it is difficult to tell a crater from a ridge, and before the age of orbiting space-probes the limb regions were poorly mapped.

There are also some really bright craters. The most brilliant of all is Aristarchus, shown in Plate III, which is always conspicuous even though it is a mere 23 miles in diameter; it can often be seen even when it is on the night side of the Moon, and lit up only by the comparatively feeble earthshine. Proclus, near the well-marked Mare Crisium, is another bright crater (Plate II). But near full phase, the lunar features of this kind are more or less overpowered by the bright streaks or rays which issue from a comparatively few craters, notably Copernicus near the junction of the Mare Nubium and the Oceanus Procellarum, and Tycho in the far south.

The rays cast no shadows; they are surface deposits, and they are well seen only under high light. When near the terminator, Tycho looks like a normal, rather bright-walled crater, 54 miles in diameter, and with a central peak. Near full, it and Copernicus are the two most striking objects on the whole of the Moon. There are numerous minor ray-centres, too, and it is notable that many of them are 'hot spots', cooling down less rapidly than their surroundings during the lunar night and during times of eclipse by the shadow of the Earth.

Of the other features of the Moon, I cannot omit mention of the valleys, such as the great gash which slices through the Alps on the border of the Mare Imbrium; the clefts or rilles, which look superficially like cracks in dried mud; the faults, of which the so-called Straight Wall is the best example: the chains of small craterlets which have been nicknamed 'strings of beads'; and the gentle swellings which we call domes. Anyone who equips himself with a telescope and an outline lunar map will soon learn to recognize the different features, and will find that learning 'lunar geography' is an enjoyable pastime. There is always something new to see.

Looking back only fifteen years or so, it is amazing to remember that our knowledge of the Moon was so slight. There was even one strange theory according to which the lunar maria were covered with layers of soft dust several kilometres deep, so that any space-craft incautious enough to land there would sink out of sight with devastating permanence. Then, in 1959, came the first Russian lunar probes, of which the most spectacular was the never-to-be-forgotten Lunik 3. In the sixties there came the American Rangers, which crash-landed on the Moon, but not before they had sent back close-range pictures during their last few minutes of flight; the Orbiters, which went round and round the Moon, sending back photographs of amazing quality; and the soft-landers, of which Russia's Luna 9 was the first. These vehicles came down gently on the Moon, so that they were able to go on transmitting after arrival. One American soft-lander, Surveyor 3, has since been visited by astronauts, and its camera brought home for analysis.

I think that when the lunar probes first flew, most people expected all the problems of the Moon to be cleared up quickly. Actually, this has not happened. The problems remain, and we have not even solved the most controversial of all: that of the origin of the Moon's craters.

On Earth, there are craters of various kinds. Some are produced by volcanic action, which is predictable enough. There are also a few craters which are unquestionably due to meteoritic impact; of these, the best-known is the Coon Butte crater in the Arizonan desert. No doubt both volcanic and impact craters are to be found on the Moon, but it is not easy to decide which process has played the major rôle.

I do not pretend to be unprejudiced. I have always thought that the principal craters are volcanic structures of the caldera type, and nothing that we have learned from the Apollo or Luna programmes has made me change my mind. In any case, the formation of a lunar crater cannot have been a violent process. When one structure interrupts another, as with Thebit (Plate IV), the wall of the broken crater remains perfect right up to the point of junction; a meteorite impact, or for that matter a tremendous volcanic outburst, would have caused obvious devastation. On the other hand, activity of the caldera type would produce just this sort of pattern, as has happened in many places on Earth.

If the craters had been produced entirely by impact, their distribution would be random; the bombardment would have been haphazard (unless, as has been commented by the German astronomer von Bülow, the impacting meteorites were highly educated!). Yet the lunar craters are not spread higgledy-piggledy all over the Moon. They form pairs, chains, groups and lines; there is nothing random about them. Patterns and alignments would be expected with craters produced by internal action. Also, it is noticeable that when one crater breaks into another, as happens on a great many occasions, it is almost always the smaller crater which intrudes into the larger. This is true in well over 99 per cent of known cases.

Possibly there is no basic difference between a large crater and a 'sea', except in size; circular maria such as Imbrium and Crisium look very much like larger versions of craters such as Ptolemæus. There is no doubt now that the seas are lava-plains, and that the lava came from inside the Moon.

Significantly, there are no major seas on the Moon's far side, though there are craters in plenty.* To me, this indicates that the Earth's gravitational pull had powerful effects on the Moon during the crater-forming period; and it is also true that most of the long chains of large craters are roughly lined up with the central meridian of the Moon as seen from Earth. If this is coincidence, it is a very peculiar one.

We must also consider the time-scale of events. In the early days of the Earth–Moon system, it is likely that the Moon rotated much more quickly than it does now; but tidal forces slowed it down until it had become captured or synchronous at 27·3 Earth-days. By the time that this had happened, a tidal bulge had been produced in the Moon and was 'frozen', so to speak, so that the Moon today is very slightly pear-shaped, with the tidal bulge facing Earthward. I suggest that most of the main surface features of large size were formed after the rotation had been captured, which would account both for their distribution with respect to the central meridian and for the fact that there are no larger maria on the far hemisphere.

The Moon is not quite so inert as many astronomers used to think. A little mild activity there lingers on. Faint, short-lived reddish patches have been seen in and near some of the craters, notably Alphonsus, one of the members of the great Ptolemæus

* I predicted this state of affairs before the ascent of the first Luniks, in 1959. Unfortunately, my original paper also contained the statement that 'it will be many years before proof or denial can be obtained'. Actually, Lunik 3 went on its journey only a few months later!

chain, and the brilliant Aristarchus (Plate IV). Amateurs have seen them often enough, but until the late 1950s their reports were officially disregarded, simply because few professional astronomers ever looked seriously at the Moon. As soon as they began to do so, they also saw the reddish patches, now known as Transient Lunar Phenomena or T.L.P.s (a term for which I believe I am responsible!). The patches are most often seen when the Moon is near perigee, and the strain upon the lunar crust is at its greatest. It is reasonable to assume that they are caused by gases escaping from below the surface, and that the Moon retains considerable internal heat, though opinions differ.

It would be misleading to compare a lunar T.L.P. with an Earth-type volcanic eruption. The two are very different both in nature and in scale; the modern Moon is a quiescent world, and violent outbursts belong to the remote past. Even a 'young' crater, such as Tycho, must surely date back to a period corresponding to the Earth's Pre-Cambrian, more than 500 million years ago. On the other hand there must equally certainly be many small impact craters, some of which must be very recent. Both vulcanism and meteoritic bombardment have been concerned in the moulding of the lunar surface.

In November 1969 the astronauts of Apollo 12, Charles Conrad and Alan Bean, deposited a seismometer on the Moon. It worked well, and signals received from it during the subsequent months confirmed that mild 'moonquakes' do take place. They are not due to impact, and are of internal origin; they too are commonest near perigee, and there is an undeniable link with the frequency of the T.L.P.s.

Artificial tremors were caused when the abandoned parts of the Apollo 12 lunar module were deliberately crashed on to the Moon. To everyone's surprise, the seismometer recorded vibrations which went on for a long time; it was even said that the Moon 'rang like a bell'. The same sort of experiment was carried out from Apollo 13, in 1970—the voyage which so nearly ended in disaster. This time the last stage of the massive rocket launcher was crashed on to the Moon, and the vibrations went on for almost three hours. Various explanations have been offered, but as yet we cannot claim to know the full answer.

Earlier, much had been heard about 'mascons', a name coined from the original term of *mass concentrations*. These are high-density areas below the crust, centred upon circular maria such as Imbrium and Serenitatis. Of course, they are not visible, but they affect the movements of probes in orbit round the Moon. Inevitably it was suggested that they were buried masses of iron, representing the original meteorites which produced the seas. This I doubt; such a mass would have to be either improbably large or improbably dense. It seems to me much more likely that the mascons have no sharp boundaries, and are merely concentrations of volcanic rock.

Perhaps it is as well to leave the argument there, if only because we have genuine hopes of resolving it during the next few years. I am well aware that the picture I have presented is one-sided, and I am ready to be proved wrong. Time will tell.

Lunar samples have been brought back, both by the Apollo astronauts and by unmanned Russian probes. Much has been learned from them. The rocks are very ancient, and date back about 4,500 million years, so that they are roughly the same age as the oldest rocks we find on Earth. They are of volcanic type, with large numbers of tiny glassy particles nicknamed 'marbles'. No meteoritic material has been found, and there is no evidence of hydrated materials (that is to say, materials which indicate the former presence of water), which disposes of the idea that the maria were once true oceans. And, of course, there is no sign of life, either past or present. As almost everyone had expected, the Moon is sterile, and has been so throughout its long history. Quarantining of the returned astronauts and samples was a wise precaution, but it was undertaken more on principle than for any other reason.

We need no longer speculate about the view from the Moon, because we have first-hand reports; and rather than give my own words, I propose to quote Neil Armstrong. Talking to me in 1970, he said:

'The sky is a deep black when viewed from the Moon ... The Earth is quite beautiful. It looks small and quite remote, but it is very blue, and covered with white lace of the clouds; the continents can be clearly seen, though they have very little colour from that distance. ... (On the Moon) you have the impression of being on a desert-like surface, with rather light-coloured hues. Yet when you look at the material from close range, as if in your hand, you find it's a charcoal grey. We had some difficulties in perception of distance. For example, our television camera we judged to be about only 50 to 60 feet away from the lunar module; yet we knew that we had pulled it out to the full extension of a 100-foot cable. Similarly, we had difficulty in guessing how far the hills out on the horizon might be. . . . (When flying over the Moon's far side) there were no observable differences in colour. The topography is the striking change. Of course, there are no seas on the far side of the Moon: it's all highlands and high mountains and big craters.'

Progress during the past few years has been remarkable. Conrad and Bean made a pin-point landing in the Oceanus Procellarum, and were able to walk across to the old automatic probe Surveyor 3, which had come down in 1967 and had sent back thousands of photographs before its power failed. Neither have the Russians been idle. They have approached the Moon in a different way—by unpiloted vehicles, controlled from Earth. Lunokhod 1, sent up in the Luna 17 probe of November 1970, marked the start of a new era. It was a 'crawler', which moved about the grounded probe, making measurements of all kinds and sending back both data and pictures. It looked like a cross between a mechanical saucepan and an antique hansom cab; but it worked, and it represented a magnificent technical triumph on the part of the Soviet team.

Though we cannot yet claim to know all about the Moon, we have learned more during the past ten years than our predecessors were able to do in ten centuries. We know that the surface is firm, and made up of lava-type rock; that there is no atmosphere, water or life; and that even the 'smoothest' parts are very rough when seen from close range. In the future, it will be the site of a full-scale scientific laboratory as well as an astronomical observatory. Neil Armstrong commented to me that 'in many ways it's more hospitable than the Antarctic might be. There are no storms, no snow, no high winds, no unpredictable weather; and as for the gravity, it's a very pleasant kind of place to work in.'

Look at the Moon now, on any clear night, and you can see the grey maria, the mountains and the craters over a distance of a quarter of a million miles. It is hard to credit that men have actually been there, and that automatic transmitters are still sending back their data from the surface. Yet with all this, the Moon has lost none of its magic.

Chapter Nine

MARS

BEYOND THE EARTH–MOON system we come to the red planet Mars, which has always had a special interest for us. Half a century ago it was thought quite possible that there might be intelligent life there, and there were many stories about 'signals from the Martians'. Even when this idea had to be given up, it was still thought that Mars could support a great deal of vegetation. This was certainly my own view; before 1965 I would have rated the chances of Martian life at well over ninety per cent.

The situation today is different, and since the first unmanned probes by-passed Mars our whole picture of the planet has been overturned. We still cannot prove that Mars is devoid of life, but the evidence, as it stands at the moment, is heavily against the existence there of any organic material. Instead of being a sort of second Earth, as had been believed, Mars has proved to be much more like the Moon.

But before going into details, a few facts and figures may be helpful.

Mars moves round the Sun at a mean distance of 141,500,000 miles. Its orbit is decidedly eccentric, and the actual distance ranges between 154 million miles at aphelion down to only 129·5 million miles at perihelion. This has a marked effect on the seasons during the 687-day long Martian year. Because the axial tilt is much the same as ours (24 degrees, as against $23\frac{1}{2}$ degrees for Earth), southern summer there occurs when Mars is at its closest to the Sun—so that the summers in the southern hemisphere are shorter and hotter than those in the north, while the winters are longer and colder. As is only to be expected, Mars is a chilly world. On a 'hot' summer day at the equator, the noon temperature may rise to 70 degrees Fahrenheit, but any Martian night is colder than a polar night on Earth, and long before sunset the thermometer will fall well below freezing-point. The axial rotation period, or 'day', is equal to 24 hours 37 minutes 22·6 seconds—a value which is known very accurately, because the surface markings are definite and permanent, and we can watch Mars spin.

The calendar is straightforward enough. In one Martian year there are 668½ Martian days. In round numbers, northern spring (southern autumn) lasts for 194 for these; northern summer (southern winter) for 177; northern autumn (southern spring) for 142, and northern winter (southern summer) for 156. Future explorers will find nothing alien in the alternation of day and night, but they will have to become used to much longer seasons.

As we noted in Chapter 2, Mars comes to opposition at intervals of around 780 days, or every other year. In the present decade, the opposition years are 1971, 1973, 1975, 1978 and 1980. (1977 is missed out because the 1978 opposition occurs in January.) The orbital eccentricity means that not all these oppositions are equally favourable, as Fig. 37 will show. In 1971 Mars came to opposition when it was also at perihelion, and the minimum distance from Earth was less than 35,000,000 miles; at the aphelic opposition of 1980 the distance will never drop below 62,900,000 miles. At its best, as in 1971, Mars can outshine every natural body in the sky apart from the Sun, the Moon and Venus. When well away from opposition, it may be no brighter

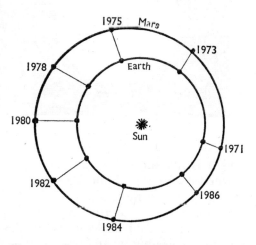

Fig. 37. Oppositions of Mars, 1971–1986. Obviously, the 1971 opposition is the most favourable; those of 1980 and 1982 the least favourable.

than magnitude 2, so that it will be comparable with the Pole Star. At such times it does look very starlike, despite its strong red colour.

Telescopically, Mars can show an appreciable phase, and can

appear the shape of the Moon a few days before or after full. For obvious reasons it never becomes a half or crescent—at least, not as seen from Earth!

Although it is so close to us on the astronomical scale, Mars is not so easy to 'observe' as might be thought. It is a small world, with a diameter of only 4,200 miles, so that in size it is intermediate between the Earth and the Moon (Fig. 38). Except when it is near opposition, only large telescopes can show much detail upon its surface (Plate VII). Even under favourable conditions, a telescopic view of Mars can never be better than a view of the Moon obtained through binoculars;

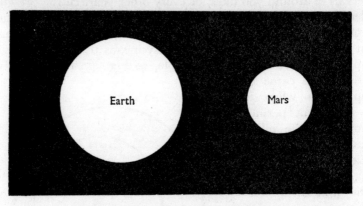

Fig. 38. Earth and Mars compared.

and it is not surprising that before the probes came to our rescue we were forced into a good deal of speculation.

The Earth, with its relatively large mass and high escape velocity, has a dense atmosphere; the puny Moon has none. Mars, then, would be expected to have a thin atmosphere, and this is exactly what is found, though we now know it to be more tenuous than we believed before 1965. There has never been any really serious suggestion that Earth-type creatures of advanced type could breathe there, and the hypothetical Martians so popular several decades ago were always assumed to be built upon a different and unfamiliar pattern.

Mars differs from Venus inasmuch as its visible markings are clear-cut and basically permanent. They were first drawn by the Dutch astronomer Christiaan Huygens (Fig. 39) as long

Fig. 39. Huygens' sketch of Mars, 1659. The
Syrtis Major is clearly recognizable.

ago as 1659, and maps of reasonable accuracy had been drawn
up by the end of the nineteenth century. The main features
were named; the nomenclature we use today is based upon that
introduced in 1877 by Giovanni Virginio Schiaparelli, the
Italian observer who will always be associated with the so-
called Martian canals—of which more anon.

Look at Mars through a large or moderate-sized telescope,
when the planet is well placed, and there will be much to see.
Even a small instrument will show the dark patches, together
with the reddish tracts and the whitish caps covering the poles
(or, more precisely, whichever pole happens to be tilted in our
direction). There is absolutely no doubt that a seasonal cycle
operates on Mars. During winter, the polar cap is large and
brilliant; with the arrival of warmer weather it shrinks, until by
midsummer it has become very small (Fig. 40). The south
polar cap has been known to vanish entirely; remember that
in the southern hemisphere the temperature-range is at its
greatest.

Also, the dark areas change. As the cap shrinks, what has
been called a 'wave of darkening' spreads equatorward from
the pole, so that the grey areas become sharper and more intense. The phenomenon is not nearly so marked as many
observers claim, but it does exist, and it has to be explained.

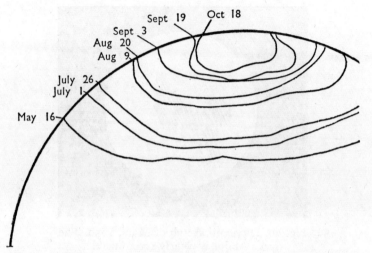

Fig. 40. Shrinkage of the south polar cap of Mars in 1956, from observations made by Patrick Moore with his 12½-in. reflector.

In the pre-Space Age period, nothing seemed easier. The polar caps were taken to be icy or snowy, and the dark areas were vegetation, which began to sprout with the arrival of moisture-laden winds. Mars is very short of water, and there can be no seas or even large lakes, so that the thin atmosphere too is desperately dry.

This dryness is unquestionable, because during winter the polar caps are very extensive—and yet release insufficient moisture to be easily detected by our Earth-based equipment. It follows that a snowy or icy cap would have to be very thin, with a depth of a few inches at most: it might be nothing more than a surface layer of hoar-frost. Any Martian vegetation would be both lowly and hardy, so that it could be no more highly-developed than terrestrial moss or lichen, though astronomers always accepted that it might be unlike any vegetation found on Earth.

Alternative theories were proposed. It was suggested, for instance, that the dark patches might be due to 'hygroscopic salts'—that is to say, salts which picked up moisture from the shrinking caps, and darkened in the process. Less inspired was the idea that the regions were nothing more nor less than ash, ejected from volcanoes which obligingly erupted at regular intervals over the Martian year. Yet on the whole the vegetation theory was favoured, if only because nobody could think of anything better.

Everything depended on the extent and composition of the Martian atmosphere. Figures published during the 1950s, mainly by Gérard de Vaucouleurs in France, indicated that the main constituent was likely to be nitrogen, with small quantities of carbon dioxide and other gases, and a vanishingly small amount of oxygen and water vapour. It was also thought that on the surface of Mars the atmospheric pressure must be of the order of 85 millibars, or about the same as the pressure in our own atmosphere at 53,000 feet above sea-level. When we remember that Mount Everest reaches upward for less than 30,000 feet, and that no mountaineer can breathe there without using an oxygen mask, it is obvious that no Earthman could survive unaided on Mars.

Clouds in the planet's atmosphere are not infrequent, and have been under observation for many years. Some of them are found at high levels above the surface, and are often called 'Martian cirrus', though they may not be composed of ice-crystals in the same way as our fleecy cirrus clouds, and rainfall on Mars is certainly unknown. The lower-lying 'yellow clouds' sometimes cover up large areas of the planet, and hide the surface features for days or even weeks at a time. It is tempting to suggest that they are due to dusty material whipped up from the reddish tracts by wind-systems of the cyclone variety, though it must be borne in mind that any wind in the tenuous Martian atmosphere will have very little force even if it is quick-moving.

This, then, was the situation before the launching of the first Martian probe. Great progress had been made in charting; features such as the dark, V-shaped Syrtis Major in the southern hemisphere, and the wedge-like Mare Acidalium in the northern, were familiar to every observer. Occasionally a dark region would spread out on to the adjacent ochre 'desert', though after a few years it usually retreated again; what could be more logical than to attribute this to the temporary spread of vegetation? The deserts themselves could not be sandy Saharas, and there was an evident lack of palm-trees, oases and camels: instead they were regarded as flattish tracts coated with coloured dusty stuff. Minerals such as felsite and limonite were likely candidates. Percival Lowell compared their appearance with that of the famous Painted Desert of Arizona.

Lowell is a character who looms large in the Martian story. In 1894 he established a major observatory at Flagstaff, in Arizona, mainly to study Mars; and he worked away diligently with his powerful 24-inch refractor up to the time of his death in 1916. He had been intrigued by the 'canals',* straight, artificial-looking streaks which had been reported in 1877 and subsequent years by Schiaparelli from Milan. As time went by, more and more people saw them, and canals became all the rage. It was claimed that they were apt to become double without the slightest warning, so that a canal might turn abruptly into a pair of strictly parallel streaks. They crossed not only the deserts, but also the dark regions; and, like everything else on Mars, they obeyed the seasonal cycle.

Lowell was nothing if not definite in his views. To him, the canals were artificial, and had been constructed by the Martians to form a planet-wide irrigation system. Every scrap of water would have to be drawn from the polar ice; dark patches where the canals intersected were centres of population, named by Lowell 'oases'. He wrote: 'That Mars is inhabited by beings of some sort or other is as certain as it is uncertain what those beings may be.'

In the age of the Mariners it is only too easy to smile at these somewhat extravagant ideas. Yet there was nothing unreasonable about them in Lowell's time, and regular canals would be difficult to explain by any natural cause. However, there were many people who could not see the canals at all. Into this category came astronomers who were using telescopes even more powerful than Lowell's 24-inch refractor at Flagstaff. Lowell further weakened his case by drawing linear features on other bodies, notably Venus and the four chief satellites of Jupiter. Even during his lifetime his views met with strong opposition, and it is probably true to say that among scientists, at least, he was always in a minority.

On the other hand, it would be dangerous to dismiss the

---

* Schiaparelli called them *canali*, which is Italian for 'channels', but everyone knows them as canals.

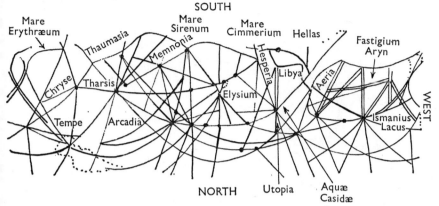

SOUTH

Mare Erythræum · Thaumasia · Mare Sirenum · Mare Cimmerium · Hellas · Fastigium Aryn · Chryse · Memnonia · Hesperia · Libya · Aeria · Tharsis · Elysium · Ismanius Lacus · WEST · Tempe · Arcadia

NORTH · Utopia · Aquæ Casidæ

Fig. 41. Network of Martian canals, according to Lowell.

canals as mere tricks of the eye. I may perhaps be allowed to cite my own observations here, because I think they are relevant even though I make no claim to being keen-eyed. I have never seen a 'Lowell-type' canal, but on the sites of some of the canals drawn on Lowell's maps I have made out broad, rather irregular streaky markings, so that the canals have a basis of reality (Fig. 41). Large telescopes are needed for this sort of work. The regular canal-patterns so often drawn by observers using small instruments are purely illusory; it is a well-known fact that the human eye does tend to join up irregular features into sharp lines. Neither have the canals ever been photographed at all convincingly.

To do Lowell justice, he never suggested that the canals could be open waterways. To be visible at all from Earth they would have to be several miles broad, and the evaporation problems alone would be crippling. It was more rational to suggest that a canal consisted of a narrow stream, possibly piped, surrounded to either side by a strip of irrigated land.

By 1940 or thereabouts most people had discarded the idea of regular channels, but the problems were still acute, particularly since it was thought that the dark regions were almost certainly due to vegetation. It was only with the flight of Mariner 4 that the first really reliable evidence came to hand.

Mariner was launched in November 1964, and by-passed Mars at about 6,000 miles on July 15 of the following year. It

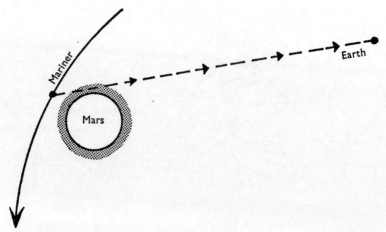

Fig. 42. The 'occultation experiment'. As the Mariner passes behind Mars as seen from Earth, there is a brief period when its signals reach us after having passed through the Martian atmosphere (shaded). In the diagram, the height of the Martian atmosphere is very much exaggerated, for the sake of clarity.

was not intended to land; it was a fly-by, equipped with recording instruments and a television camera. It was designed to analyse the planet's atmosphere, and to send back close-range pictures of the surface. Both aims were carried through, and the results caused an astronomical sensation of the first magnitude.

The photographs were spectacular. Instead of having a flat surface, Mars proved to be scarred with craters basically similar to those of the Moon. The general aspect was clear; craters existed both on the ochre tracts and on the dark regions.

Secondly, it was found that the atmosphere of Mars is much less dense than had been thought. The experiment designed for this particular investigation was remarkably ingenious. The probe passed behind Mars, and was hidden or occulted; just before occultation, its radio signals came to us by way of the atmosphere round the planet's limb, as shown in Fig. 42. As the signals went through the Martian atmosphere they were affected and distorted, which enabled the American researchers to work out approximate values for the density and even the composition. At once they found that the surface pressure must be far less than the accepted value of 85 millibars.

The next step came in 1969, a few days after Armstrong and Aldrin made their landing on the Moon. Mariners 6 and 7 went past Mars, and repeated the experiments, with more reliable results. Again craters were shown, and the occultation procedure was carried out. We can now say, with confidence, that the ground pressure on Mars is no more than 6 to 7 millibars, equivalent to that in our air at well above 120,000 feet above sea-level (Fig. 43). Moreover, the main constituent is not nitrogen, but carbon dioxide.

This is even more significant than it appears at first sight. The Martian atmosphere is useless so far as Earth-type creatures are concerned, and it is so inefficient at blanketing in the Sun's day-time warmth that the nights are very cold indeed. Worse, it seems that the atmosphere is unable to shield the surface from lethal short-wave radiations from the Sun and space. If so, then Mars may be radiation-soaked and sterile.

It follows, too, that the polar caps are not made up of H_2O in any form. They are more likely to be solid carbon dioxide—an idea which had been proposed even in Lowell's time, but which had met with only sporadic support.

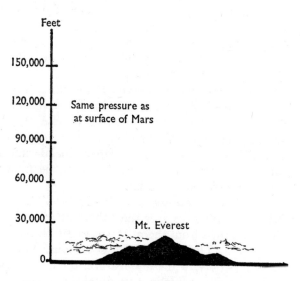

Fig. 43. Pressure of the Martian atmosphere, compared with that in our own air.

How do these findings affect the dark areas? If not due to vegetation, then what are they?

Here we have to admit that we are uncertain, and it seems that we know much less about Mars than we thought we did ten years ago. The problem has been further complicated by some important work carried out by R. A. Wells, who has used spectroscopic methods to study the amount of carbon dioxide over various parts of Mars. It is logical to assume that these amounts will be least over elevated areas, greatest over valleys (just as there is less air over Tibet than there is over England). Wells has found that the dark areas are not necessarily low-lying basins, as used to be thought. Some of them are high; others are not. Thus the Syrtis Major is a rather broad, lofty area which slopes off steeply to both sides, while the desert of Amazonis is a sunken basin larger than our Mediterranean. The dark Mare Sirenum and the ochre Memnonia, both shown in the map on page 111 (Fig. 44), make up another depression. Mars is not a world with a flat landscape; it is decidedly rugged.

Shifting dust is one suggestion made to account for the dark regions and their seasonal cycle. The wind-systems on Mars should be fairly regular, and dust-particles of different sizes, and hence of different reflecting powers, will be deposited and then swept away again according to the season. Yet the theory is neither wholly plausible nor wholly understandable, and we must await the results from further probes.

We may, however, have the answer to the canal puzzle. The main streaks charted by Lowell and his followers seem to be made up either of mountainous uplands or else of chains of roughly-aligned craters. As had been fully expected, the Mariners have shown no features with artificial aspect. The brilliant-brained Martians must, with regret, be finally relegated to the realm of science fiction.

The Mariner 6 and 7 photographs (*see* Plates VII, VIII, IX, X, and XI) were much clearer than those of Mariner 4, and more could be learned from them. It was found that not all the surface is cratered. There are 'chaotic' regions, with tangled ridges and valleys only, while one feature, the circular whitish plain of Hellas (easily visible in a small telescope when Mars is near opposition) proved to be blank. Either no features had

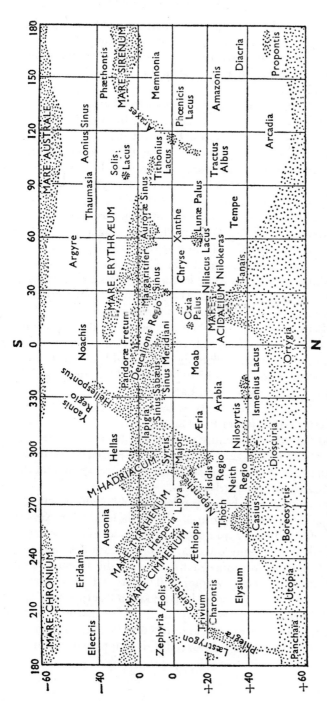

Fig. 44. MAP OF MARS, drawn by Patrick Moore from personal observations in 1963 with his 12½-in. and 8½-in. reflectors. The projection is a Mercator, and the nomenclature follows that drawn up by the International Astronomical Union.

ever existed inside it, or else the interior has been levelled in some way, presumably by volcanic re-melting. Craters near the south pole were seen to be covered with the characteristic whitish deposit, and there was strong evidence of erosion here and there.

Inevitably there came arguments as to whether the Martian craters were due to volcanic action, meteoritic impact, or a combination of both processes. I do not intend to discuss the problem here at any length, because the answer should be found within the next few years. There can be no reasonable doubt that the craters on Mars are of lunar type; and once we have found out which force played the dominant rôle on the Moon, we shall have the answer for Mars as well. In maintaining that vulcanism has been the main agent, though not the sole one, I realize that I am courting criticism!

One exciting theory, due to B. C. Murray in the United States, is that activity on Mars is still in progress. He has suggested that heat is being generated below the planet's crust, and that this leads to the melting of otherwise permanently-frozen soil just below the surface; when this happens, the surface itself crumples and suffers partial collapse, so that any existing craters are destroyed. Certainly this would account for the chaotic areas. Also, it would explain the blankness of Hellas, which the impact theory can hardly do.

Let us stress that Mariners 4, 6 and 7 between them have covered only a fraction of the Martian surface, and it is always dangerous to draw too many conclusions from evidence which is so limited. Yet the general trend is clear, particularly since the pictures are so good; the smallest structures shown on them are no more than 300 yards across.

The next step will be to send up probes which will orbit Mars, sending back photographs of the whole surface. Next will come a soft landing by an automatic vehicle: then, perhaps, a probe which will land on Mars, collect samples and come home, as the Russian Lunas have done from the Moon. Quarantining of the returned samples will be a real necessity. The Moon is undoubtedly sterile, but we cannot yet be sure about Mars, and to run risks with any alien life-form would be foolhardy in the extreme. Great care must also be taken not to carry Earth contamination to Mars; if this happened, future ex-

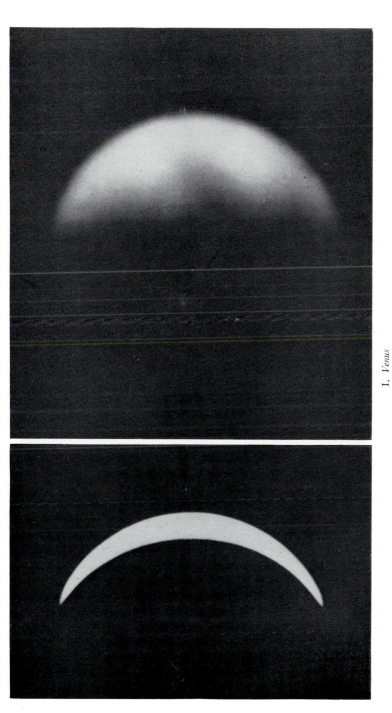

I. *Venus*

(*left*) Photograph with the 200in. Palomar reflector. (*right*) Photograph in blue light: 100in. reflector, Mount Wilson.

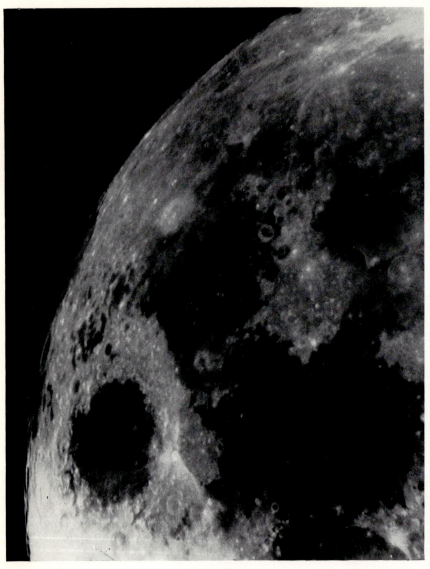

II. *The Moon*

Photograph by Commander H. R. Hatfield (12in. reflector), 1967 September 18. The Mare Crisium is shown to the lower left; the Mare Tranquillitatis occupies the lower centre and right.

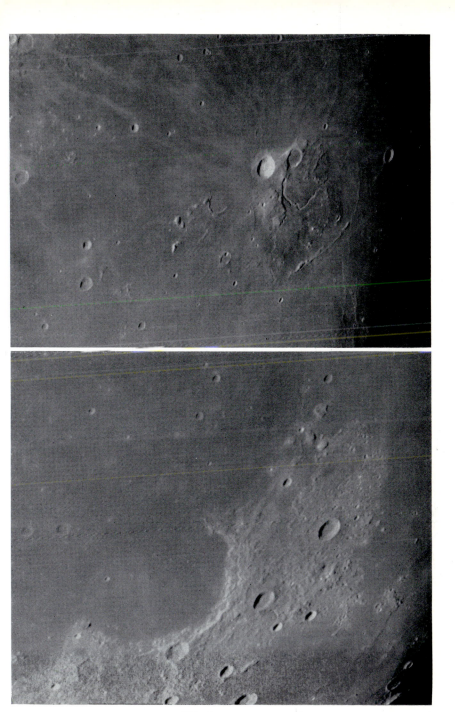

III. *Lunar scenery*

(*above*) The area of Aristarchus, with the winding valley extending from the companion crater Herodotus.
(*below*) The Sinus Iridum (Bay of Rainbows), which leads off the Mare Imbrium. The landing site of Lunokhod 1 was in this area of the Mare Imbrium, just off the top right.
Photographs by Commander H. R. Hatfield (12 in. reflector).

IV. *Lunar scenery*:

(*left*) The area of the Straight Wall, visible as a dark line. The three craters to the lower left are Ptolemaeus (partly shown), Alphonsus and Arzachel. (*right*) A crowded upland area in the south part of the Moon. The prominent deep crater, centre right, is the ray-crater Tycho, 54 miles in diameter. The rays are not shown here, as the photograph was taken under conditions of low illumination.

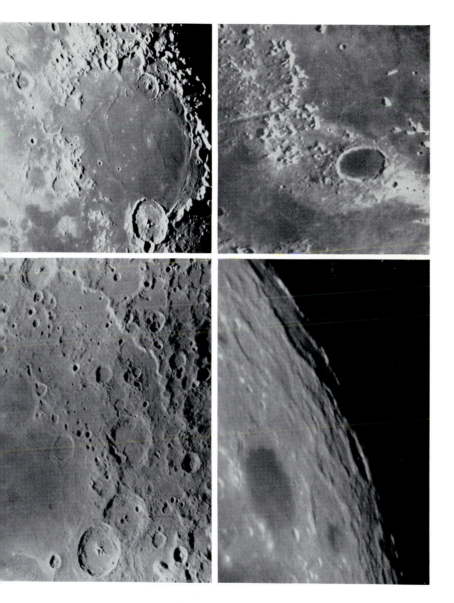

V. *Lunar scenery*

(*above left*) Gassendi and the Mare Humorum (Commander H. R. Hatfield).
(*above right*) Plato and the Alpine Valley (R. Smith).
(*below left*) Part of the Mare Nectaris, with the Altai Scarp and the chain of craters
Theophilus-Cyrillus-Catharina (Commander H. R. Hatfield).
(*below right*) Grimaldi and Riccioli area on the Moon's limb (W. Rippengale, 10in. reflector).

VI. *The Earth*

A photograph taken from Apollo 8 by Colonel Frank Borman from the neighbourhood of the Moon: December 1968.

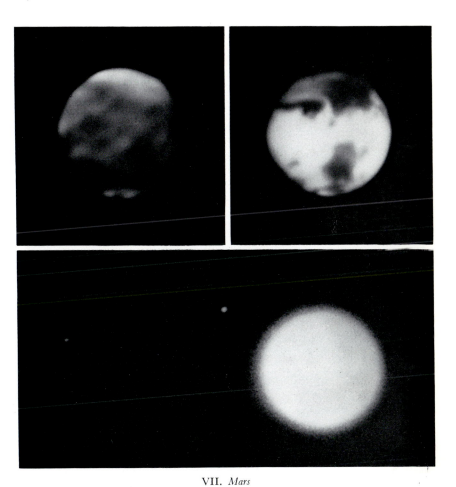

VII. *Mars*

(*above*) Mars, in blue and red light: 200in. reflector, Palomar.
(*below*) Mars and its satellites: G. P. Kuiper, 82in. reflector (McDonald Observatory),
1956 September 14. The image of Mars is necessarily over-exposed. Phobos and Deimos
appear as spots of light to the left of the planet.

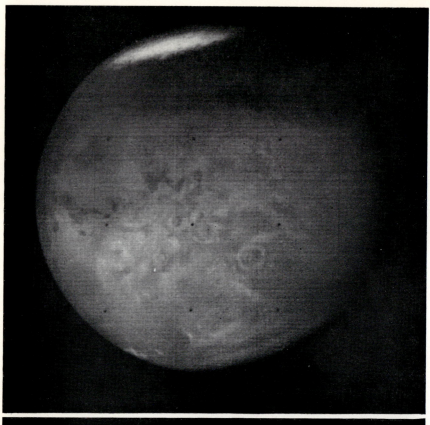

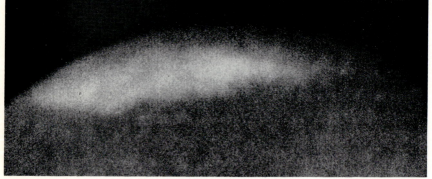

VIII. *Mars, from Mariner 7; 1969*

(*above*) View at 293,200 miles from Mars; 1969 August 4, 10.28 G.M.T. Central meridian
224.6° E. Prominent in the picture are the bright, ring-shaped Nix Olympica and the bright
streaks of Tharsis-Candor. Mare Sirenum is to upper right. South is at the top.

(*below*) The South Polar Cap from Mariner 7. The boundary of the cap is quite sharp, but
irregular in shape.

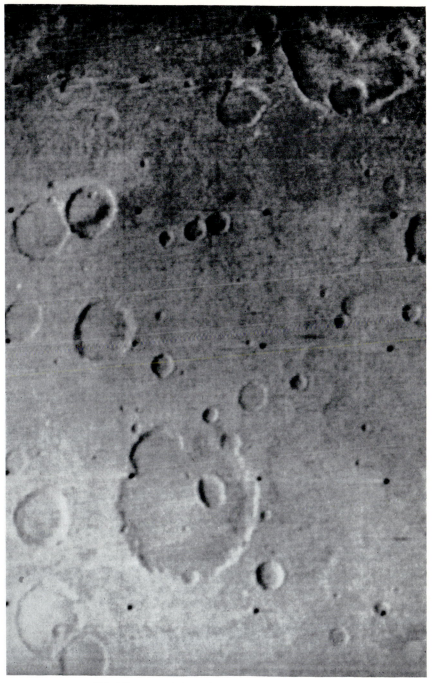

IX. *The craters of Mars, from Mariner 6: 1969 July 30*

The small black dots are part of the reference points within the television system. The distance of Mars at this time was 2,150 miles from the space-craft. The photograph was taken through a red filter during the 20-minute period surrounding Mariner's closest approach to Mars.

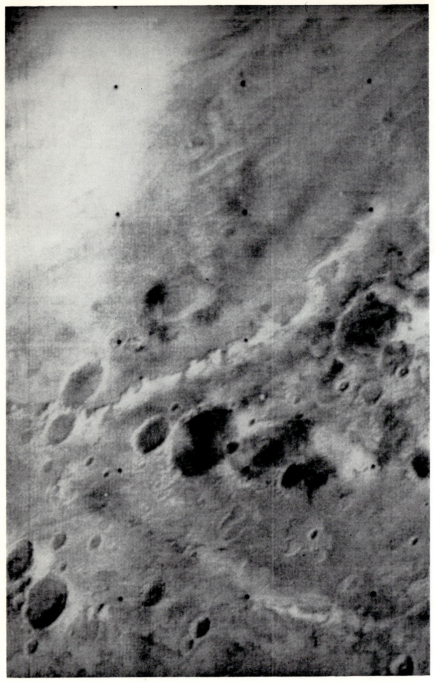

X. *The Martian south polar cap region, from Mariner 7*

This is one of the most spectacular of the space-craft pictures; the whitish deposit shows up clearly. When the photograph was taken, it was not far from sunset over this part of Mars.

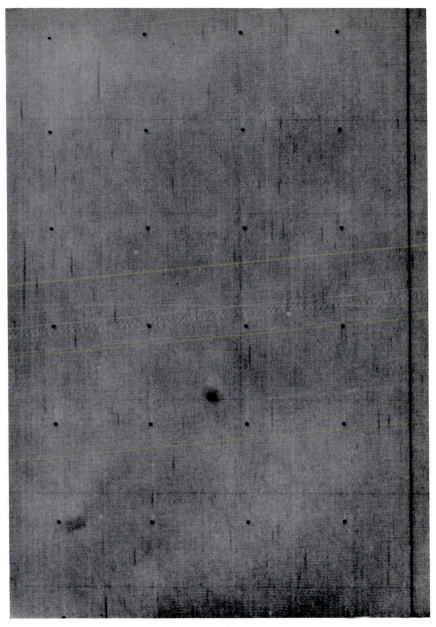

XI. *Hellas, the circular plain on Mars*

Photographed from Mariner 7 through a red filter. The picture covers an area of 400 × 550 miles. The black dots are part of the TV reference system. It is obvious that the floor of Hellas is almost devoid of detail; no craters appear. There is no chance that this blankness is due to local obscurations in the Martian atmosphere; it is unquestionably real.

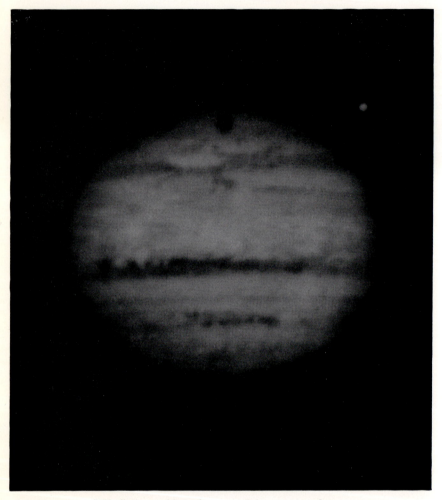

XII *Jupiter : red light photograph, Palomar 200in. reflector*

The third satellite Ganymede is shown, together with its shadow in the south polar region of Jupiter.

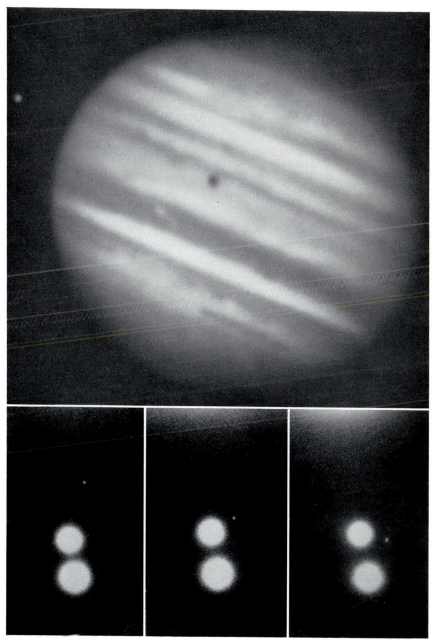

XIII. *Jupiter and Satellites*

(*above*) Jupiter: Palomar 200in. reflector. Note the satellite shadow near the centre of the disk. (*below*) Three satellites of Jupiter: G. P. Kuiper, 1955 January 30, 82in. reflector (McDonald Observatory). The large lower satellite is Ganymede, the upper Europa; the tiny fifth satellite Amalthea can be seen on all three pictures as a spot of light.

XIV. *Saturn*

(*above*) Saturn: photographed with the 100in. reflector at Mount Wilson. The ring system is wide open.

(*below*) Saturn and its inner satellites: G. P. Kuiper, 1948 March 24 (82in. reflector, McDonald Observatory). The image of Saturn is necessarily over-exposed.

XV. *Saturn close to the Moon*

Photographed by Commander H. R. Hatfield (12in. reflector) during the close conjunction of 1967.

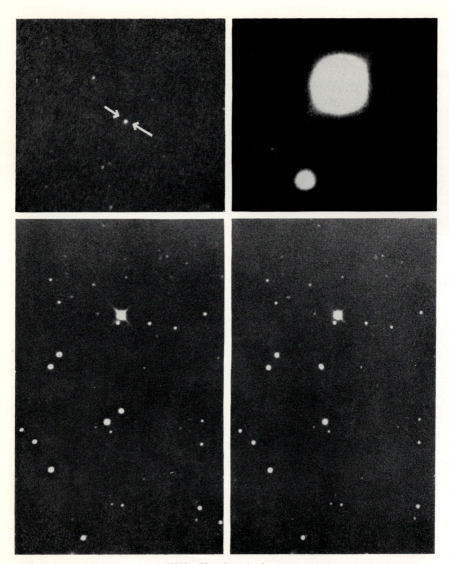

XVI. *The telescopic planets*

(*above left*) The asteroid Vesta, indicated by the arrows; photograph by Frank Acfield, 10in. reflector.

(*above right*) Neptune, with its satellite Triton: G. P. Kuiper, 82in. reflector, 1949 February 24.

(*lower pair*) Pluto, photographed by Kuiper on 1950 January 24 and 25 with the McDonald 82in. reflector. Note the movement of Pluto, which is to the lower centre of the picture in the left-hand photograph, but in the right-hand view has moved closer to the brightest star.

plorers would be unable to study the planet in its 'mint condition', so to speak. We have no guarantee that terrestrial bacteria would not spread and multiply quickly even in the hostile Martian environment.

The orbiting probes and pioneer soft-landers should have done their work by 1975. Manned flight to Mars must surely await the development of nuclear-powered rockets, but the Americans plan to send a crew there by 1990, and there is no reason to suppose that they will fail.

About the internal constitution of Mars we know little as yet. The overall density of the globe is less than that of the Earth, and any heavy, iron-rich core must be relatively smaller; one estimate for its diameter is 1,000 miles. So far we have no evidence of any magnetic field.

Lastly, something should be said about the two Martian satellites, Phobos and Deimos, both of which were discovered by Asaph Hall in 1877. They are extremely small. Phobos was shown clearly on one of the Mariner 7 photographs, and is shaped rather like a potato, $14\frac{1}{2}$ miles long and $11\frac{1}{2}$ miles wide; Deimos is somewhat smaller (Fig. 45). They are faint objects, well beyond the range of small telescopes. I have seen them clearly with a 15-inch reflector, but they are extremely elusive, and it is small wonder that they were overlooked before Hall's systematic search revealed them.

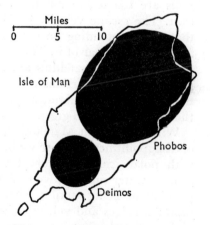

Fig. 45. Phobos and Deimos, compared with the Isle of Man.

Phobos is peculiar in that it moves at about only 3,600 miles from the surface of Mars, and completes one revolution in 7 hours 39 minutes. This is much less than the Martian 'day' of 24 hours 37 minutes—a case unique in the Solar System, so far as we know. To an observer on Mars, Phobos would rise in the west and set in the east $4\frac{1}{4}$ hours later, during which time it would go through more than half its cycle of phases from new to full. The interval between successive risings would be just over 11 hours. Deimos, over 12,000 miles from the planet's surface, has a period of 30¼ hours, so that it would remain above the observer's horizon for two and a half Martian days at a time.

Neither would provide much illumination at night. Phobos would have about one-third the apparent diameter of the Moon as seen from Earth; Deimos only one-ninth. For long periods when above the horizon they would be eclipsed by the shadow of Mars, and in the case of Deimos the phases would be hard to distinguish with the naked eye. They could cause no solar eclipses, but they would often pass in transit across the face of the Sun; Phobos would do so 1,300 times each Martian year, taking less than twenty seconds to pass right over the disk. Because their orbits lie in the plane of the planet's equator, they would never rise from high latitudes on the surface of Mars.

Certainly these dwarf attendants are unlike our Moon, and it has been suggested that they may be captured asteroids. They are also very dark, probably because their gravitational pulls are too feeble for them to collect any interplanetary dust. The escape velocity of Phobos is a mere 30 m.p.h., so that anyone standing on the satellite would be able to throw a cricket-ball clear of it. The strange movements have led to a theory that both satellites are hollow space-stations, built by past or present Martians for reasons of their own. This is intriguing, but, alas, about as plausible as the age-old idea that the Moon is made of green cheese.

When we look at Mars today, it is difficult not to feel a sense of disappointment. Where we had hoped to find a living world, with polar ice, vegetation tracts and perhaps even underground water supplies, we find a planet which gives every impression of being sterile; where there are huge craters and jumbled ridges and valleys, with polar caps made up of solid carbon dioxide. Yet we cannot be sure that conditions are as depressing as they look, and the probes of the 1970s may provide us with some major sensations. In any case, the Red Planet remains one of the most fascinating of all our neighbour worlds.

Chapter Ten

THE MINOR PLANETS

ANY PLAN OF the Solar System shows the huge gap between the orbits of Mars, outermost of the terrestrial planets, and Jupiter, first of the giants. Kepler was struck by it, and suspected that it might contain an extra member of the planetary system. He went so far as to write: 'Between Mars and Jupiter I put a planet.'

He knew that even if such a planet existed, it could not be large, as otherwise it would have been visible with the naked eye. For over a century the problem was forgotten, but it was revived in 1772 by a German astronomer named Johann Bode. Somewhat earlier, another German, Titius, had discovered a curious numerical relationship between the distances of the planets from the Sun. Bode publicized it, and nowadays it is known—rather unfairly!—as Bode's Law. Whether it has any real significance is a matter for debate. It may be pure coincidence, but it is certainly peculiar.

Take the numbers, 0, 3, 6, 12, 24, 48, 96, 192 and 384, each of which (after 3) is double its predecessor. Now add 4 to each, giving: 4, 7, 10, 16, 28, 52, 100, 196, 388. Taking the Earth's distance from the Sun as 10, this series of numbers gives the distances of the remaining planets, to scale, with remarkable accuracy, as is shown in the following table:

Planet	Distance by Bode's Law	Actual Distance
Mercury	4	3·9
Venus	7	7·2
The Earth	10	10
Mars	16	15·2
—	28	—
Jupiter	52	52·0
Saturn	100	95·4
Uranus	196	191·8
Neptune	—	300·7
Pluto	388	394·6

The three outer planets, Uranus, Neptune and Pluto were not known when Titius worked out the relationship; but when Uranus was discovered, in 1781, it fitted excellently into the general scheme. Neptune, admittedly, is a 'problem child'. According to Bode's Law it ought not to be there, and the last figure (388) corresponds well enough to the actual mean distance of Pluto. Yet at the time of its discovery the Law seemed very precise—except for the missing planet corresponding to figure 28 in the scale.

In 1800 six astronomers assembled at the little German town of Lilienthal, where the hard-working amateur Johann Schröter had his observatory. They nicknamed themselves the 'celestial police', and determined to make a serious effort to track down the missing planet. With Schröter as President, and Baron Franz Xavier von Zach as Secretary, they worked out a scheme in which each member would be responsible for a particular section of the ecliptic—that is to say the region where the planet would probably lie, assuming that it existed at all.

A plan of this sort takes some time to bring into working order, and before Schröter's 'police' were fully organized they were forestalled. Piazzi, director of the Sicilian observatory of Palermo, was compiling a star catalogue, and on January 1, 1801—the first day of the new century—he picked up a starlike object which behaved in a most unstarlike manner inasmuch as it showed perceptible motion over a period of even a few hours. Piazzi at first took it to be a tail-less comet, but he went so far as to write to von Zach; evidently he had his suspicions. Postal services were slow and unreliable (as, indeed, they are today!) and by the time that von Zach received Piazzi's letter, the moving body had been lost in the evening twilight.

Fortunately Piazzi had made enough observations to enable an orbit to be worked out, and the great mathematician Gauss, who tackled the calculations, soon saw that he was dealing with a planet rather than a comet. It was re-detected exactly a year after its original discovery, and Piazzi named it Ceres in honour of the patron goddess of Sicily. It was found to have a distance of 27·7 on the Bode scale, which corresponded most satisfactorily with the expected 28; the real distance from the Sun amounted to 257,000,000 miles, with a period of 4·6 years. The Solar System was apparently complete.

Yet the 'celestial police' were not so sure. Ceres, with its diameter of less than 500 miles, seemed hardly worthy to be ranked with the other planets, and so Schröter, von Zach and their colleagues pressed on with their search. It came as no real surprise when one of them, Dr. Heinrich Olbers, picked up a second small planet in March 1802. Pallas, as it was named, was so like Ceres in size and distance that Olbers believed the two to have been formed from one larger body which had met with disaster. The idea was attractive; if there were two fragments, there might be more—and two additional planets came to light within the next five years. Juno was discovered by Karl Harding in 1804, and Vesta by Olbers in 1807.

Juno and Vesta resembled the first two members of the group, and the four became generally known as the Minor Planets or asteroids. No more seemed to be forthcoming, and the 'celestial police' disbanded in 1815. Schröter himself died in the following year.

There matters rested until 1830, when a Prussian amateur named Hencke took up the problem and began a systematic search for new asteroids. Alone and unaided, he worked away for fifteen years, and at last had his reward—in the shape of a fifth minor planet, now named Astræa, circling the Sun at a distance slightly greater than Vesta's, slightly less than Juno's. However, Astræa was considerably fainter than the original four, and its estimated diameter is only just over 100 miles.

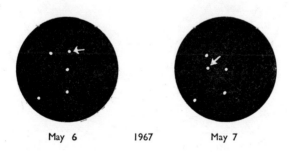

May 6   1967   May 7

Fig. 46. Movement of Vesta in 24 hours: Patrick Moore, 1967 (3-in. refractor).

Even the enthusiastic Hencke would have been surprised to learn that his discovery was a mere prelude to thousands more. He himself found another asteroid, Hebe, in 1847: in the same year Hind, in London, discovered Iris and Flora; 1848 and 1849 yielded one asteroid each, and since then every year has produced its quota. By 1870 the total number of known asteroids was 109 and twenty years later it had grown to 300. Then, in 1891, Max Wolf of Heidelberg introduced a new method which led to a startling increase in numbers.

Wolf's method was a photographic one. If a camera is adjusted so as to follow the ordinary stars in their movement across the sky, an asteroid will show up as a streak across the plate—because an asteroid moves against the stars quickly enough for its shift to be noticeable with a time-exposure of

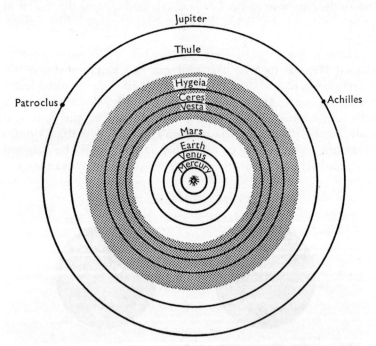

Fig. 47. The main asteroid belt, shaded. It extends from inside the orbit of Vesta out to that of Hygeia. Thule, which was for many years the most remote known asteroid, lies outside the main swarm. Achilles and Patroclus are Trojans, moving in the same orbit as Jupiter.

only an hour or two. If I set my camera to photograph a garden, and then walk in front of the lens during a time-exposure, I will appear as a blur, because of my movement. The asteroid will not blur, since it is a hard, sharp point of light, but its motion will certainly betray it (Fig. 46).

The Wolf method was almost embarrassingly successful, and the numbers of known asteroids increased by leaps and bounds. Wolf was personally responsible for adding over a hundred, and by the beginning of 1971 there were more than 1,750 asteroids with properly worked-out paths. At least a thousand more have been found on photographic plates without having been under observation for long enough to have their orbits computed (Fig. 47).

It cannot be said that the asteroids proved to be popular members of the Solar System. Plates exposed for quite different reasons were often found to be swarming with short tracks, and the irritating little planets complicated star-counts and similar work to such an extent that German astronomers, who nobly established a computing centre to try to keep pace with them, started referring to the 'Kleineplanetenplage' (minor planet pest). One American observer went so far as to call them 'vermin of the skies'.

Another difficulty was to find names for them. As the numbers grew, dignified mythological names such as Psyche, Thetis, Melpomene and Circe began to give out. The first departure was with No. 25, Phocæa, named by Benjamin Valz after a seaport in Ionia; the next case was 45 Eugenia, commemorating the wife of the French emperor Napoleon III. Some of the later names are tongue-twisting; we have 678 Fredegundis, 989 Schwassmannia, 1259 Ogylla, 1286 Banachiewicza, and so on. No. 724 is Hapag, the initials of a German navigation line, the Hamburg Amerika Paketfahrt Aktien Gesellschaft, while 674 is Ekard, the word 'Drake' spelled backwards—it was named by two members of Drake University in America. 518 Halawe takes its name from the favourite dessert of its discoverer, R. S. Dugan, who had developed a liking for the Arab sweet *halawe*. It is perhaps fortunate that 1581 Abanderada (a mob-leader who carries a banner) is not too close to 1197 Rhodesia: and I particularly like 1372 Haremari, which was named by Karl Reinmuth in recognition of the female members (or harem!) of the astronomical institute of which he was Director. One name was actually sold. Asteroid 250 was discovered by Palisa, Director of the Vienna Observatory, who wanted to raise funds for an eclipse expedition. He therefore announced that he would sell the honour of naming his asteroid at a price of £50. The offer was taken up by Baron Albert von Rothschild, who chose his wife's name, Bettina. There are various other amusing stories, and it is with regret that I refrain from pursuing the subject further!

Ceres, Pallas, Juno and Vesta are known as the 'Big Four'. In fact Juno, about 150 miles in diameter, is surpassed by some

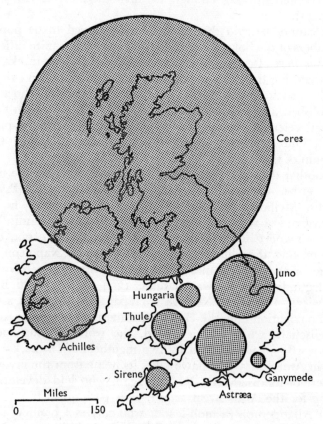

Fig. 48. Sizes of some of the asteroids, compared with Great Britain.

others, notably No. 10, Hygeia, which may be as much as 220 miles across. Asteroid diameters are extremely difficult to measure, and the present results are probably not very inaccurate, but there seems little doubt that Ceres—427 miles—is the largest of the swarm (Fig. 48). Next comes Vesta, with 370 miles. This is a new measurement, made in 1970 by D. A. Allen, using infra-red techniques; it puts Vesta ahead of Pallas (280 miles). Vesta is closer to the Sun than Ceres, and therefore also closer to us; its mean distance from the Sun is 219,300,000 miles, and it is the only minor planet to be visible with the naked eye. Keen-sighted people can find it without optical aid, and I have seen it myself without much difficulty, though I doubt whether I should have noticed it if I had not known it to be there. Ceres, Pallas, Juno and other asteroids are within the range of binoculars, though of course they look exactly like stars, and are identifiable only because of their movement from night to night.

No asteroid is massive enough to retain any trace of atmosphere, and most of them are mere lumps of material, probably not even approximately spherical. The total membership of the swarm is uncertain; one American estimate gives 44,000, while the Russian astronomers tend to think that 100,000 may be nearer the mark. Yet if all the asteroids were combined into one body, they would not add up to a planet nearly as massive as the Moon.

One or two of the early members have unusual paths. That of Pallas is inclined by 34° 48', so that it can move well away from the ecliptic; in 1971 it spent part of its time in the southern constellation of Eridanus, the River. No. 279, Thule, is unusually remote, since it circles the Sun at a mean distance of almost 400 million miles. Toward the end of the last century it was noticed that the asteroids tend to fall into groups; the effect is quite genuine, and is due to the tremendous disturbing influence of giant Jupiter. However, at that time all the known asteroids kept strictly to the gap between the orbits of Jupiter and Mars, and nobody was prepared for the odd behaviour of No. 433, Eros, which was discovered in 1898 by Witt at Berlin.

Eros is small, and never bright. Its strangeness lies in its orbit, which swings it well inside the main group. It has an eccentric path, and its aphelion takes it beyond Mars, but at perihelion it can come within fourteen million miles of the Earth. Close approaches are rare; the last was in 1931, when the minimum distance from us was 17,000,000 miles, and the next is due in 1975.

When at its nearest Eros can have its position measured very accurately. Its distance can be found, and this gives a key to the whole scale of the Solar System. Hundreds of photographic measurements made of it in 1931 were analysed by Sir Harold Spencer Jones, who arrived at a value of 93,003,000 miles for the length of the astronomical unit or Earth–Sun distance. This is now known to be too large, and much better methods

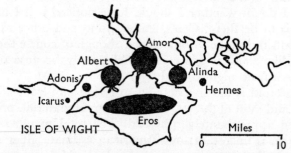

Fig. 49. Sizes of some of the 'Earth-grazer' asteroids, compared with the Isle of Wight.

have been developed, so that Eros has lost its importance, and studies of it in 1975 will be less extensive. However, it has played its part, and astronomers of a few decades ago admitted that it provided some compensation for its irritating fellows.*

In 1931 it was found that Eros shows light-variations in an average period of about five hours. Since no planet or asteroid has any light of its own, the only explanation is that Eros is irregular in shape, and is rotating. This was confirmed visually by van den Bos, who saw the asteroid oval at times. It seems to be 15 to 18 miles long and 4 to 5 wide, so that it does truly resemble a piece of cosmic débris (Fig. 49).

For some time Eros was thought to be unique (Fig. 50), but in 1911 Palisa, at Vienna, picked up a tiny body which can

* It is easy to see why the asteroids are often regarded with distaste. Some decades ago S. B. Nicholson was using the 100-inch Mount Wilson reflector to search for faint new satellites of Jupiter. Altogether his photographic plates recorded 32 unexpected asteroids, all of which had to be eliminated as possible Jovian moons, and which wasted an incredible amount of time.

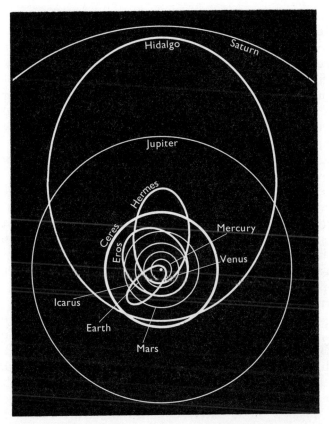

Fig. 50. Orbits of some of the asteroids.

approach the Earth to within 20,000,000 miles, though its orbit is so eccentric that its aphelion distance is almost as great as that of remote Thule. It was numbered 719, and named Albert. Unfortunately it is only about 3 miles in diameter, and after its brief visit in 1911 it vanished into the distance; so far it has not been re-discovered. A tiny body such as Albert is susceptible to even the slightest perturbing force, and recovery is bound to be largely a matter of luck. At the moment we simply do not know where it is, though it may be found again one day.

887 Alinda, discovered by Wolf in 1918, and also 1036

Ganymede,* found by W. Baade in 1924, are other asteroids with orbits of the Albert type; but all these, Eros included, were outdone by Amor and Apollo, the two 'earth-grazers' of 1932.

1221 Amor, discovered by the Belgian astronomer Delporte, is a full 5 miles across. It came within 10,000,000 miles of the Earth, and was under observation for long enough for its orbit to be reliably determined. Amor has a period of 975 days, and after it had been twice round the Sun unseen it was picked up again in 1940. Since then it has not been lost.

However, Amor's reign as a record-holder was brief. Apollo, discovered by Reinmuth at Heidelberg later in 1932, approached the Earth to within 7,000,000 miles. At perihelion it is only 59,000,000 miles from the Sun—closer-in than the Earth or Venus—and so it can play some strange tricks. Like Mars and the other outer planets, it is best seen at opposition; but it can also pass through inferior conjunction, and it can even transit the Sun's disk, though it is so small that it could not possibly be observed during a transit. Unhappily it, like Albert, has been lost.

This was also the fate of Adonis, discovered by Delporte in 1936, which veered past us at only 1,300,000 miles; at perihelion it approaches the orbit of Mercury. A year later, Reinmuth found an even more interesting earth-grazer, subsequently named Hermes. Even smaller than Adonis, with an estimated diameter of only a mile, it passed by at a distance of only 485,000 miles, barely double that of the Moon. It is theoretically possible for it to come still closer, actually passing between the Moon and the Earth.

Needless to say, astronomers were not in the least alarmed by this celestial visitor. Hermes may have been very close in the astronomical sense, but there was no danger of a collision; in fact, the chances of our being hit by an earth-grazer are many millions to one against. If we reduce the Earth in scale to a 12-inch globe, Hermes may be compared with a speck of dust passing several feet away.

It is of course true that if a collision occurred, the damage would be widespread. In 1908 a meteorite with a diameter of

* This is an unsuitable name, since the third satellite of Jupiter is also called Ganymede. Asteroid 1036 is rather larger than the other members of its group; the estimated diameter is 20 miles.

perhaps a quarter of a mile hit Siberia, devastating an area of many square miles—and there is no difference, in name, between a large meteorite and a small asteroid.* When the Hermes story was made known, January 1938, the Press made the most of it, and the headlines of national papers on 10 January were highly sensational. 'World Disaster Missed by Five Hours,' was one example. 'Scientists Watch a Planet Hurtling Earthward.' Yet Hermes, interesting though it may have been, was certainly no threat.

Mars, closer to the main swarm that we are, also has its visitors. 1009 Sirene can pass within 5,000,000 miles of it, and doubtless there are many other small asteroids which go even closer in.

For many years Thule was thought to be the outermost member of the asteroid swarm, but in 1906 Max Wolf at Heidelberg detected No. 588 Achilles, which was obviously more remote. In fact, it was found to move in the same orbit as Jupiter, so providing mathematicians with an interesting demonstration of what is termed a Lagrangian point.

As long ago as 1772, the famous French mathematician Lagrange had called attention to the special 'problem of three bodies', which arises when a massive planet and a tiny asteroid move round the Sun in the same plane, in circular orbits and with equal periods. Lagrange found that if the bodies are 60 degrees apart, they will always remain 60 degrees apart. Achilles behaved in just this way. Subsequently other similar asteroids were found, and were given the names of combatants in the war between Greece and Troy, so that nowadays these remote asteroids are known collectively as the Trojans.

Sixteen Trojans have been discovered, but two have been lost again because they were not observed for long enough to have reliable orbits worked out. The remaining fourteen are split into two groups on opposite sides of Jupiter. Sixty degrees ahead of Jupiter lie 1404 Ajax, 659 Nestor, 1647 Menelaus, 624 Hector, 911 Agamemnon, 1143 Odysseus, 1437 Diomedes, 588 Achilles and 1583 Antilochus; sixty degrees behind came 1173 Anchises, 1208 Troilus, 617 Patroclus, 1172 Æneas, and 884 Priamus.

* Some Russian astronomers believe that the Siberian object may have been the nucleus of a small comet. However, the principles involved are just the same.

The Trojans do not, of course, keep strictly 60 degrees ahead of or behind Jupiter, because we are dealing with elliptical orbits, and there are various other factors to be taken into account—notably perturbations caused by Saturn. Diomedes, for instance, may go as far as 40 degrees beyond the 60-degree point on the side away from Jupiter, and 24 degrees from it on the side toward Jupiter. All the members of the group are faint; the largest, Hector, is around 150 miles in diameter, while Menelaus may be no more than a dozen. There can be no doubt that other Trojans exist, in addition to the two which were found and then lost again.

Jupiter has seven small satellites which move at great distances from the planet, and it has been suggested that these are nothing more than captured Trojans. On the other hand, it is equally possible that the Trojans themselves are ex-satellites of Jupiter which somehow 'got away'. So far we do not have enough information to decide one way or the other.

It was thought that the Trojans must mark the extreme outer limit of the asteroid swarm, but this did not prove to be so. Hidalgo, discovered by W. Baade in 1920, is not a Trojan, but has a most extraordinary orbit which carries it from inside the main group out to almost as far as Saturn. Its period is fourteen years, and its path is so eccentric that it moves more in the manner of a comet than an asteroid. It was carefully photographed with the 100-inch reflector at Mount Wilson, but always showed up as a sharp point of light, devoid of any trace of the fuzziness which betrays a comet. Therefore, we are bound to include Hidalgo in the minor planet family, though with the inner feeling that it is something of a black sheep!

The movements of the more normal asteroids are not without interest. As we have seen, the minor planets tend to fall into well-defined groups or families, separated by almost unpopulated zones known as Kirkwood gaps in honour of their discoverer, Daniel Kirkwood. These barren regions are due to planetary perturbations (mainly caused by Jupiter). Five of the families include numerous asteroids whose orbits are so alike that they seem to have had a common origin. This brings us back to Olbers' theory of a disrupted planet which used to circle the Sun between the paths of Mars and Jupiter.

For many years the Olbers hypothesis was discounted, but

now it seems to be coming back into favour. According to the
Dutch astronomer Oort, the original planet exploded; the
fragments thrown into nearly circular orbits provided the
asteroids and meteorites, while those with more elliptical orbits
were so violently perturbed by Jupiter and the other giant
planets that some were driven out of the Solar System altogether, while the remainder formed an outer cloud of comets.
It has also been suggested that there were two or more original
planets, which broke up by collision. However, there is no
general agreement as to what happened, and it may well be
that the asteroids (and meteorites) never formed one larger
body, in which case they presumably represent the débris left
over after the main planets were born.

Before leaving these miniature worlds, let us turn to perhaps
the most remarkable asteroid of all—Icarus, discovered by
Baade in June 1949 (Fig. 51). When found, it was about
8,000,000 miles away
from the Earth, and it can
never approach us closer
than half this distance, so
that it is not an earth-
grazer in the sense that
Adonis and Hermes are. It
is unique inasmuch that
at perihelion it passes the
Sun at only 18,000,000
miles, much closer-in than
Mercury, so that it is violently heated. At perihelion
it recedes to 183,000,000

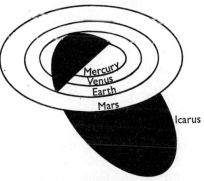

Fig. 51. The highly eccentric and
inclined orbit of Icarus.

miles from the Sun, well beyond the orbit of Mars. The temperature-range during the Icarian 'year' of 409 Earth-days is
obviously tremendous.

Icarus last paid us a visit in the summer of 1968. It came to its
minimum distance, but even so it was sixteen times as far away
from us as the Moon, and there was absolutely no chance that it
would come any closer. For some strange reason, reports
originating from Australia caused a certain amount of uneasi-
ness among non-astronomers, and it was even suggested in the
popular Press that there might be a head-on collision. The

report was not officially denied for some time, simply because astronomers did not think it worth bothering about—and before a statement was made, a question about it had been asked in the House of Commons!

However, Icarus by-passed us strictly on schedule; it became bright enough to be visible in a moderate telescope as a dot of light, moving at almost one degree per hour. (The apparent diameter of the full moon is half a degree.) Slight changes in magnitude indicated a rotation period of between one and two hours, and it also seems that Icarus is slightly bluer than the average asteroid. On June 13, 1968 it was picked up by the radar equipment at the Lincoln Laboratory at Lexington, in America: later it was also detected by radar elsewhere in the United States, and contact with it was kept up until June 16.

Icarus is named after the mythological youth who took to the air by using artificial wings, and who met an untimely death when he flew too near the Sun, so that the wax fastening his wings melted. The strange little world can by-pass all the inner planets; for instance, on May 1, 1968 it was within ten million miles of Mercury. Studies of its movements are theoretically important, and luckily the orbit of Icarus is so well known that there is no fear of its being lost.

Within the next few decades, no doubt some of the asteroids will be photographed from close range, by camera-carrying rocket probes. Whether landings will ever be made on them is problematical, but there is no valid reason why not—even though it would be very over-optimistic to hope to find any useful or valuable materials there. Science-fiction writers (and also a few official science writers, who should certainly know better) have suggested fixing motors to asteroids and towing them into convenient orbits. Alas, the amount of force needed would be so great that nothing of the kind is even remotely possible. Asteroid beacons, as navigational aids, are by no means improbable in the future. There may also be automatic recording stations on them—even upon Icarus, alternately scorched and frozen during its strange, wildly extreme 'year'.

To the ordinary observer, the asteroids are of little interest; they show no measurable disks, and even when found they seem hardly worth the trouble spent on the search. Yet they are by no means without their value; and the remarkable orbits of Eros, Achilles, Hidalgo, Icarus and their kind show that it is unjust to dismiss all the minor planets as mere 'vermin of the skies'.

Chapter Eleven

JUPITER

FAR BEYOND THE main asteroid zone we come to Jupiter, giant of the Solar System, and appropriately named after the King of the Gods (Fig. 52). It moves at a mean distance of 483,000,000 miles from the Sun, and has a 'year' almost twelve times as long as ours. There can be no proper seasons, since the axis of rotation is almost perpendicular to the orbital plane; but seasons would in any case be meaningless on a world so cold as Jupiter.

The orbital velocity is less than half that of the Earth, so that the synodic period is only 399 days, and Jupiter comes to opposition every year—though during the early 1970s it is in the southern part of the sky, rather low down as seen from Europe. All the same, it is very brilliant. It is outshone only by Venus and, very occasionally, by Mars.

Jupiter is a good reflector of sunlight, but its brilliance is due mainly to its tremendous size. Measured through the equator, the diameter is 88,700 miles, but the polar diameter is less,

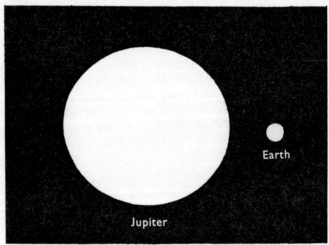

Fig. 52. Jupiter and Earth compared.

because the globe is appreciably flattened; any small telescope will show the effect.

The flattening is caused by Jupiter's rapid axial rotation. It has a shorter 'day' than any other planet; at the equator, the period is only 9 hours 50½ minutes, and particles there are being whirled round at 28,000 m.p.h. Away from the equatorial zone the rotation period is five minutes longer, though there are many variations, and different surface features have rotation periods of their own. No solid body could spin in this fashion. Therefore, Jupiter cannot be 'solid' in the usual sense of the word; its outer layers, at least, are made up of gas. It follows that Jupiter is of comparatively low density, and though it has a volume over 1,300 times greater than that of the Earth it is only 318 times as massive. This is very considerable for a planet; indeed, it exceeds the combined masses of all the other planets together, and it has been said that the Solar System is made up of 'the Sun, Jupiter, and various minor bodies'. On the other hand, Jupiter is very much less massive than even a lightweight star. The distinction is important when we come to consider what kind of a world it really is.

First, it is cold—at its surface. The temperature there is around −120 degrees Centigrade. Yet Jupiter is probably very hot inside, and according to one estimate the temperature near the centre of the globe is as much as half a million degrees. Also, Jupiter seems to emit rather more energy into space than it ought to do in view of the amount it receives from the Sun. Is it possible, then, that Jupiter is itself a 'minor sun'? This idea was popular until the 1920s, and was not entirely dispelled by the discovery that the outer clouds are so chilly.

However, theoretical considerations rule it out, and show that there is an essential difference between a planet and a star. The distinction hinges chiefly upon mass—and remember, it would take over a thousand Jupiters to make up one body as massive as the Sun.

A star is formed out of a cloud of dust and gas. The material is drawn together by gravitational force; as the particles move inward towards the centre of the contracting cloud, they collide with each other, and the general temperature is raised. When it reaches 14 million degrees Centigrade, nuclear reactions begin, and the star starts to shine. There is a definite critical temperature at which these nuclear reactions are triggered off. Different authorities give rather different values; but it is quite definite that a body with a mass as low (by stellar standards!) as that of Jupiter will not reach a core-temperature high enough for the process to begin. Indeed, the mass is too low by a factor of at least 30, and Jupiter is definitely not a star of any kind.

Yet we must find some reason for the half-million degree temperature near the centre of its globe. Radioactive heating is one possible answer. It is also possible that Jupiter is slowly shrinking, and releasing gravitational energy. This is by no means out of the question; and if the globe shrank under its own weight by as little as a millimetre per year, we could account for the extra energy which is being sent out. However, there is no proof, and there is always the chance that Jupiter's interior is much cooler than half a million degrees.

As yet, our knowledge of the make-up of the giant planet is far from complete. By using spectroscopic analysis, we can find out the nature of the upper gases, and it has been found that hydrogen compounds are plentiful; but when we 'go inside' we are reduced to pure theory, and every investigating astronomer seems to have his own ideas.

As long ago as 1872 some prominent dark lines were found in the spectrum of Jupiter (*see* Plate XIII). In 1907 Lowell, at Flagstaff, managed to take some excellent photographs of them, but it was not until much later that they were correctly interpreted. Rupert Wildt, in the United States, showed that they were due to ammonia and methane, both of which are hydrogen compounds.* This led to a widespread impression that the outer layers of Jupiter must be made up chiefly of ammonia and methane, with only small quantities of other substances. Actually, this is not so. In 1963 H. Spinrad and L. Trafton, again in the United States, published an analysis in which they suggested that ammonia and methane together made up only one per cent of the Jovian 'atmosphere'. Neon accounted for three per cent and helium for 36 per cent; the remaining 60 per cent was composed of hydrogen. Unfor-

* Ammonia is made up of hydrogen and nitrogen (chemical formula NH_3); methane, better known by its common name of marsh-gas, is carbon and hydrogen (CH_4). Under terrestrial conditions—that is to say, mixed with oxygen—methane is dangerously explosive. Miners dread it, and call it 'fire-damp'.

tunately hydrogen and helium are very shy of showing themselves spectroscopically under the conditions to be expected on
Jupiter, while ammonia and methane are highly obtrusive.

Not everyone accepts the Spinrad–Trafton analysis; some
authorities increase the percentage of hydrogen to around 80.
However, there is no doubt that hydrogen is abundant, and
this is no surprise at all, since in the universe as a whole the
amount of hydrogen exceeds those of all the other elements put
together. Rather than dismiss Jupiter as freakish, we must
explain why the inner planets, including the Earth, are so
hydrogen-poor. Nobody has yet managed to give a good
answer, though it has been suggested that in the early days of
the Solar System the intense solar radiation drove the light
hydrogen and helium atoms outward; at greater distances,
such as that of Jupiter, the process would be much less efficient.

Moreover Jupiter, with its great mass, has a very high escape
velocity: 37 miles per second. This means that it was capable of
holding on to all its original hydrogen, whereas the Earth was
unable to do so.

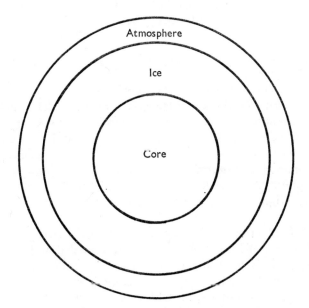

Fig. 53. The structure of Jupiter, according to
Wildt's theory.

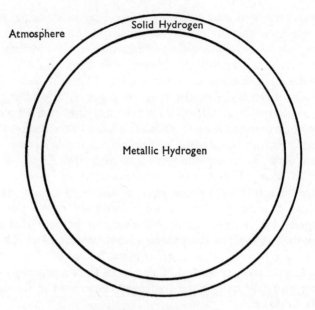

Atmosphere   Solid Hydrogen

Metallic Hydrogen

Fig. 54. The structure of Jupiter, according to
Ramsey's theory.

Whatever may be the basic cause, Jupiter now has a hydrogen-rich outer layer. What we have to decide is whether the
rest of the globe is of similar composition.

Wildt did not think so. He produced a model which was
widely accepted for many years, according to which Jupiter is
'layered' (Fig. 53). There is a rocky, metallic core 37,000 miles
in diameter, overlaid by an ice shell 17,000 miles thick; above
this comes the atmosphere. Conditions at the bottom of the
atmosphere, 8,000 miles below the visible surface, are strange
indeed. High pressure makes a gas behave in a most un-gaslike
manner.

The first serious challenge to the Wildt model came from
W. Ramsey, of Manchester, who maintained that Jupiter was
more likely to be made up chiefly of hydrogen all the way
through its globe (Fig. 54). On Ramsey's model we begin, as
before, with an outer atmosphere of hydrogen, together with
hydrogen compounds such as ammonia and methane, plus
some helium. Below the visible surface the pressure rises

rapidly and the density of the hydrogen increases, so that it takes on the characteristics of a solid. At a depth of 2,000 miles, the pressure is 200,000 times that of the Earth's air at sea-level, and the solid hydrogen has a density one-third that of water.

At 5,000 miles there is a sudden change. The pressure has grown to 80,000 atmospheres, and instead of being merely a highly-compressed solid the hydrogen starts to behave like a metal. Metallic hydrogen is more easily compressible than the more ordinary solid hydrogen, and at the centre of the planet the density is well over three times that of water.

On Ramsey's pattern there is a core of metallic hydrogen 76,000 miles in diameter, accounting for over 90 per cent of the total mass of the planet. Above comes a shell of solid hydrogen 5,000 miles deep, and on the outside of this there is the relatively shallow atmosphere, ending in the visible surface shown in our telescopes.

A variant of this idea is due to P. Peebles, who believes that there is a core of dense metallic elements and rocky silicates, above which come, in order, a layer of metallic hydrogen, a layer of hydrogen in solid and/or liquid form, and the outer atmosphere. The pressure at the bottom of the Jovian atmosphere would be 200,000 times that of the Earth's air at sea-level, and the temperature would be around 2,000 degrees.

At the moment we do not know which—if any!—of these models is correct. No doubt more will be learned from the fly-by probes due to be launched within the next year or two. Meanwhile, we can say with absolute confidence that Jupiter is quite unlike the Earth, and also unlike the Sun or any other star. Before turning to the composition of the atmosphere, however, we must digress to say something about the telescopic appearance. Undoubtedly Jupiter is one of the most rewarding of all the planets from an observational point of view, mainly because it is always changing, and one can never be sure what is going to happen next.

The most prominent markings on the yellowish, flattened disk are the cloud belts, which run parallel to the planet's equator. They are always very much in evidence, and generally speaking their latitudes do not change much. Between the belts are the lighter regions known as zones. The accepted nomenclature for Jupiter is given in Fig. 55; I have followed

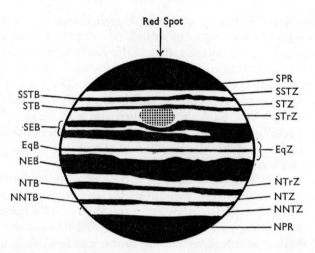

Fig. 55. The belts and zones of Jupiter. N = north; S = south; T = temperate; Tr = tropical; Eq = equatorial; P = polar; R = region; Z = zone; B = belt.

the standard astronomical practice of putting south at the top—though I fear that the American space-planners will alter this when the first probes are launched, as has already happened with the Moon and Mars!

Though the main belts are always on view, they are not constant either in intensity or in breadth. For many years now the North Equatorial Belt (NEB) has been the most conspicuous of all the features, but it has not always been so, and may not retain its pre-eminence indefinitely. Then there are, of course, other markings—notably the Great Red Spot, which will be described below.

Just to show the types of variations which take place on Jupiter, I give here some brief extracts from my own observational notebook. The instrument used was, in general, my 12½-inch reflector (together, in 1965–8, with the 10-inch refractor at Armagh Observatory). Of course, a much smaller telescope will show the main details; a 3-inch refractor will show quite a number of belts as well as the Red Spot.

1951. NEB dark and broad; SEB faint and narrow; STB prominent. Red Spot not conspicuous.

1952. NEB dark and broad; SEB not so dark, but much broader than in 1951. STB has faded. In general, a rather inactive disk.

1953–4. NEB broad, dark and double. SEB broad and double, but not so dark. EqZ very bright at times.

1955. NEB prominent; both SEB and STB narrow. Inactive disk.

1956. A striking change! NEB broad and double, but so was the SEB, sometimes the equal of the NEB. STB thin and narrow.

1957. Main belts now the NEB and STB; the SEB thin and narrow again, with the Red Spot coming back into prominence.

1958. Much the same as in 1957, but with rather more activity in the EqZ. Red Spot strongly coloured. Just before Jupiter vanished into the evening twilight, the SEB increased in breadth and darkness.

1959. NEB broad and dark; SEB usually darker and broader than STB; Red Spot not in evidence. The chief peculiarity was the abnormally strong yellowish hue of the EqZ. I first noted it on March 28.

1960. NEB broad and dark, with irregular borders; SEB prominent and double; STB also quite prominent. I saw the Red Spot only once with certainty, and there was virtually no colour in it, though other observers were no doubt more successful.

1961. An interesting year, since for some time the whole of the EqZ was dusky, and contained some very large white spots. The STB was quite prominent, but inactive; the Red Spot was evident again.

1962. An extraordinary opposition. The two equatorial belts merged into a 'solid wedge' of yellowness, in which were some white spots. The view was utterly unlike anything I had seen before, since I began observing in 1934. The STB was present, and so was the Red Spot—particularly later in the year, when the yellowness of the EqZ was less marked.

1963. The equatorial belts reappeared in recognizable form, and the abnormal yellowness had gone, but the whole EqZ was in a state of upheaval, with various white ovals and much fine structure. Red Spot reasonably prominent. STB conspicuous and regular; STZ bright. At times the equatorial belts almost merged again.

1964. Signs of a reversion to 'normal', but the SEB was at times the equal of the NEB, and the STB was very obscure. Both equatorial belts were very broad. The Red Spot was visible, but not so prominent as it can sometimes be.

1965–6. Much more normal. NEB its usual broad, dark self; SEB broad, dark and often double, though inferior to the NEB; STB regular and quite prominent; Red Spot more conspicuous. Considerable detail seen in the EqZ, and the EqB was recorded on a number of occasions.

1966–7. EqZ back to its normal brightness and colour, with a prominent, complex NEB; a double, much less intense SEB, and a rather uneven STB. Red Spot prominent. At one stage I saw an unusual darkening in the STrZ which looked as if it might be interesting, but it did not persist. The SPZ became dusky at times.

1968. Both equatorial belts prominent and double; STB also prominent; many white spots in the STZ; Red Spot conspicuous. A generally active disk. Occasionally the SEB was actually darker and broader than the NEB, though more often it was less so.

1969. The NEB regained its pre-eminence, and the SEB was not conspicuous at first, though it was variable in intensity. The yellow cast of the EqZ returned, though not so violently as in 1962, and the belts were always identifiable individually. The EqB was sometimes visible, and the Red Spot was prominent. Less activity in the STZ region.

1970. Again the STB gained precedence over the SEB, though both were less conspicuous than the NEB. The NEB and the STB dominated the disk. The Red Spot was prominent, but less strongly coloured than in, say, 1968.

I do not for one moment claim that these results are exhaustive, but they do at least represent the work of a single observer, and they may serve to give an overall indication of the kinds of changes which go on. Many observers in past years have recorded other colours—greens, yellows and even blues and violets. I can only say that I have never seen them myself, and they seem to have been noted less often during the past few decades. Whether the colours on the disk really are less pronounced today, or whether the older observers tended to exaggerate faint hues, is not clear.

Though the latitudes of the features do not change much, the longitudes do. Jupiter, as we have seen, is a quick spinner, and it does not rotate as a solid body would do; different regions have different periods. The so-called System I—that is to say, the area bounded by the north edge of the South Equatorial Belt and the south edge of the North Equatorial Belt, as shown in Fig. 55—has an average period of 9 hours $50\frac{1}{2}$ minutes; the rest of the planet (System II) takes five minutes longer. There is, in fact, a quickly-moving current along the equator. Yet these figures are only a mean; special features, such as the Red Spot, have periods of their own, and drift about in longitude relative to their surroundings. For some years prior to 1969, for instance, the longitude of the Red Spot was increasing fairly steadily.

It is not hard to measure the rotation periods of special features on Jupiter. Because the spin is so quick, a few minutes' observation will show that the surface details are being carried round the planet—from left to right with an ordinary inverted telescopic image. What has to be done is to time the moment when the object reaches the central meridian. Fig. 56 shows

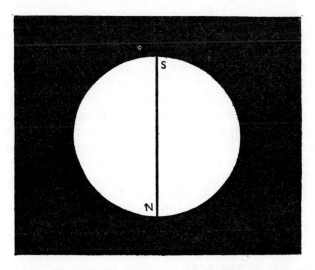

Fig. 56. Central meridian of Jupiter.

what is meant; the central meridian is easily located, because
of the flattening of the disk, and 'transits' of the surface features
can be timed to the nearest minute.* Obviously, successive
transits of the same feature will give its rotation period. (In
practice it is seldom possible to obtain two transits of the same
object in one night, because the interval is over 9¾ hours; I have
done it on suitable occasions, when Jupiter is high in the northern hemisphere, but it is not easy. However, the transits need
not be of successive rotations.)

Once the transits have been timed, reference to a set of
tables will yield the rotation period. I do not propose to give
the mathematics of it here, because the method can be found
in *Practical Amateur Astronomy* (*see* Appendix), but nobody need
be frightened; the only requirement is that the observer
should be able to add!

What exactly are the belts?

The conventional answer is: 'Crystals of ammonia', and this
seems reasonable enough. Certainly they seem to overlie the
bright zones, though even this has been occasionally challenged.
The structure of the Jovian atmosphere is still rather uncertain,
and various models have been proposed. One of those, due to
R. Gallet, gives various layers; starting from the top of the
visible cloud layer, we have, in order (i) ammonia crystal
clouds, (ii) ammonia droplet clouds, (iii) unsaturated ammonia
vapour, (iv) ice crystal clouds, (v) water droplet clouds and
(vi) unsaturated water vapour, beneath which comes the true
surface of the planet. Unfortunately we cannot be sure that
there is any definite 'surface', though there is no doubt that in
Jupiter's atmosphere the most plentiful gas is hydrogen.

Observers soon find that most of the interesting activity takes
place near the equator and in the south tropical and south
temperate zones. Not surprisingly, the poles are less turbulent,
and in general the northern hemisphere is the calmer of the
two. This may well be due to the presence in the south of one
truly remarkable feature: the Great Red Spot.

The Spot first became really conspicuous in 1878, but it was
by no means new. Hooke, contemporary of Isaac Newton,
apparently saw it in 1664; Cassini, first Director of the Paris

* This use of the term 'transit' has nothing to do with Jupiter's position in
the sky.

Observatory, drew it in 1665, and there are suggestions that it was seen even earlier. It is extremely long-lived, and must be something more than a mere 'cloud'.

In 1878 the Spot became brick-red in colour, and measured 30,000 miles long by 7,000 wide, so that its surface area was equal to that of the Earth. After 1882 it faded somewhat, and since then it has even vanished completely at times; but it always returns, and its site is marked by indications of what is termed the 'Red Spot Hollow', looking like a large bay in the South Equatorial Belt. Between 1965 and 1971 it was very prominent, with a strong pinkish-red colour which could hardly be overlooked even by the most casual telescopic observer.

Because it drifts about in longitude, it cannot be attached to a solid surface—and suggestions that it might be the top of a volcano have little to recommend them! The redness certainly does not indicate heat, and is presumably due to some exceptional chemical composition.

There seem to be only two plausible explanations for the Red Spot. One was supported by the late Bertrand Peek, an English amateur who became one of the most skilled of all observers of Jupiter. Peek regarded the Spot as a solid body floating in the outer atmosphere of the planet, and pointed out that its variations in prominence could be due to changes in level: if the Spot sank it would be covered up, while when it rose to the top of the cloud-layer it would be conspicuous. Calculations showed that the total change in level need not be more than about seven miles.

To explain the cause of this rising and falling, Peek drew a comparison with the well-known experiment of immersing an egg in a solution of salt and water. If the solution is densest near the bottom of the jar, as will probably be the case, the egg will float at a level determined by the density of the liquid. Add some more salt, thereby increasing the density of the liquid, and the egg will rise to the top of the jar. Therefore, a slight increase in the density of Jupiter's atmosphere in the region of the Spot would force it upward.

This sounds logical enough, though nobody has put forward any really sound model for the composition of the Spot itself, and its colour remains a mystery. However, there is another possibility. If Jupiter has a fairly well-defined surface below the clouds, we may expect strong winds there (or, more accurately, strong currents). If there is some large obstruction on the surface, the atmosphere might circulate round it, and above the obstruction there would be a column of stagnant gas, known as a Taylor column. If so, then the Red Spot might be nothing more than the top of this column.

As yet we cannot decide between these two theories; my preference for Peek's 'floating island' is personal only. Whatever may be its true nature, the Red Spot is unique. No other spot has lasted for nearly so long, and most are much smaller and short-lived.

Associated with the Red Spot in the days when I first began studying Jupiter was a most interesting feature known as the South Tropical Disturbance. It was of about the same latitude of the Spot, and had a slightly shorter rotation period, so that every few years it caught up the Spot and passed it. While this was going on there were marked interactions between the Disturbance and the Spot; the Disturbance was accelerated, and as it passed by it seemed to drag the Spot along with it for several thousands of miles. When the Disturbance had gone on its way, the Spot drifted back to its original position. The Disturbance was first reported in 1901; it was last observed in 1941, and seems to have gone for good. Now and then there are indications of a revival (I saw something of the kind in 1967), but on the whole it seems that these features are impermanent, of lesser importance than the original Disturbance, and not true returns of it.

In 1954 there came a new development in Jupiter research. B. Burke and K. Franklin, in America, discovered that the planet is a source of radio emission at dekametre wavelengths. (One dekametre is equal to ten metres.) The discovery was unexpected, and was also quite accidental, but it proved to be of great importance. Since then, it has been found that the radio emissions occur in bursts which are violent, erratic and irregular.

Let it be stressed at once that there was never any suggestion that the radio waves from Jupiter could be anything but natural! After all, both light-waves and radio waves are electromagnetic phenomena; the only essential difference is in wavelength. But the emissions from Jupiter were very much of a puzzle, and still remain so.

It was logical to suppose that they might be associated with some of the prominent features on the visible disk, notably the Red Spot and some of the white spots which are so often seen in the South Temperate Zone. At first there did seem to be a correlation, but studies carried out over a longer period showed that there is no real connection between radio bursts and visible features.* Thunderstorm activity in the Jovian atmosphere was proposed as an explanation, but has now been rejected simply because of the tremendous amount of energy involved. No doubt magnetic phenomena play a part in the production of the bursts, and it is possible that there may be a periodicity linked with the position in orbit of Io, Jupiter's innermost large satellite. But all in all, a full explanation must await the results to be sent back from future fly-by probes. As well as the erratic dekametre bursts, there are steadier radio emissions which indicate that Jupiter has a very strong magnetic field, together with radiation belts of the same basic type as the Van Allen zones surrounding the Earth.

Finally in this brief survey of Jupiter, we come to the satellite system, which includes twelve members—four large, eight very small.

In 1609 Galileo, using his home-made telescope, saw that Jupiter was attended by four starlike objects which soon proved to be satellites. Simon Marius, another early telescopic observer, saw them at about the same time; indeed, there ensued one of those childish squabbles about priority which bedevil science now and then. Marius named the four attendants Io, Europa, Ganymede and Callisto. The names were not officially recognized until much later, but by now they have come into general use.

All four satellites can be seen with any small telescope, or, for that matter, with good binoculars. Really keen-sighted persons can catch sight of them with the naked eye, though I certainly cannot do so myself. They are of planetary dimensions; Io and Europa are roughly the size of our Moon, while Ganymede

* I carried out a long series of observations in this programme, and the work was interesting even though it led in the end to a negative result—as so often happens in astronomy.

and Callisto are decidedly larger. Their diameters have been measured from time to time, but the results do not agree well, though they are of the right order. It used to be thought that Callisto was slightly larger than Ganymede, though other recent measurements make it smaller. Certainly Ganymede is the brightest of the family, and it is comparable in size with Mercury, though its density is lower and it is less massive.

The orbits of all four 'Galileans' lie in much the same plane, so that telescopically they tend to appear in a straight line. Various interesting phenomena are associated with them. They may be eclipsed by Jupiter's shadow; they may transit the disk (yet another use of the word 'transit'!) or be occulted by it, and their shadows also may be seen crossing the planet. All these phenomena are listed in yearly almanacs, and with a little practice it is not hard to tell one satellite from another.

The four Galileans are not alike. Io and Europa are more dense than our Moon, and are fairly reflective; with Io, G. P. Kuiper considers that an 'excess of metals' is present, possibly covered with a layer of oxide smoke. Europa has an even higher albedo, but is the least massive of the four.

Large telescopes can show surface markings, and maps have been made of all the Galileans, but the charts are of a preliminary nature only. J. Mulliney and W. McCall, at Allegheny in the United States, have claimed that moderate-sized telescopes will show the markings. Of course this is not so, as any experienced observer will verify. Really powerful equipment is needed.

Ganymede and Callisto are bigger than our Moon, but much less dense, so that they seem to resemble planets in size but satellites in mass. Ganymede has an escape velocity of 1·7 miles per second; no atmosphere has been detected, despite the most careful searches. It has been suggested that Ganymede may have a rocky core coated with a thick layer of ice or even solid carbon dioxide, but this is pure speculation.

Callisto is about the equal of Ganymede in size, but its density is only twice that of water, and it has a much lower reflecting power, so that it is actually the faintest of the Galileans. Owing to its low escape velocity, it is unlikely to have retained any vestige of atmosphere.

The transits, shadow-transits, eclipses and occultations of the four satellites are fascinating to watch, and each body has its own way of behaving. During transits across the disk of Jupiter, Io and Europa are generally hard to find except when close to the limb, whereas Ganymede and Callisto, which are less reflective, tend to show up as greyish spots. The shadows are always black. Mutual phenomena between the satellites can also occur; for instance, Io and Europa may be eclipsed by the shadow of Ganymede, and one satellite may occult another. The first reliable observation of this sort seems to have been made by Peek on May 22, 1926, when he watched an eclipse of Io by the shadow of Europa.

The eclipses of the Galilean satellites by the shadow of Jupiter itself are visible with a small telescope. In 1675 the Danish astronomer Ole Rømer found that the calculated eclipse times did not always agree with observation, and he attributed this to the fact that light travels at a finite velocity; remember, Jupiter is not always at the same distance from us, and is in motion relative to the Earth. After elaborate calculations, Rømer arrived at a value for the velocity of light which was close to the true value of 186,000 miles per second.

The remaining eight satellites of Jupiter are small and faint. Number 5, discovered by E. E. Barnard in 1892 and named Amalthea, is closer to the planet than any of the Galileans; it is about 112,000 miles from the centre of Jupiter, and has a revolution period of only twelve hours. It is subject to tremendous gravitational strain, and may be egg-shaped, though there is no proof; photographs from probes may well tell us. The diameter cannot be more than 150 miles or so, and it is beyond the range of average amateur telescopes.

The outer satellites are generally known by numbers, but this is a bad system: No. 10, for instance, is closer to Jupiter than No. 9. In view of the desperate efforts to find names for every tiny asteroid, this failure to give names to the members of Jupiter's family seems rather remiss. Some years ago, Brian Marsden (then of Britain, now carrying out researches in America) suggested that No. 6 should be known as Hestia; 7 as Hera; 8 as Poseidon; 9 as Hades; 10 as Demeter; 11 as Pan, and 12 as Adrastea (Fig. 57).

All the outer satellites are small. Hestia may be 100 miles across, the others less than 50. Hestia, Hera and Demeter lie at

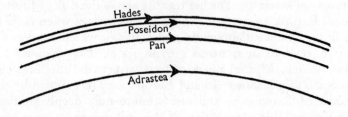

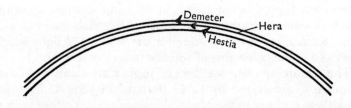

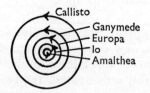

Fig. 57. Orbits of the satellites of Jupiter. It must be remembered that owing to the perturbations by the Sun, the orbits of the outer small satellites are by no means so regular as shown here.

roughly 7,000,000 miles from Jupiter, and have periods of around 270 days; Adrastea, Pan, Poseidon and Hades are between 13 and 15 million miles out, and take between 630 and 760 days to go once round Jupiter. Moreover, all four travel the 'wrong way'—east to west instead of west to east.

With small, remote satellites of this kind, the perturbations in their orbits due to the Sun are very great. The paths of the satellites round Jupiter are not even approximately circular, and change with every revolution, so that the movements are hard to predict with real accuracy; indeed, Poseidon was lost for some time after 1941, and was not picked up again until 1955. There is nothing illogical in the idea that junior satellites of such a kind are captured Trojans.

As yet, we have been able to do no more than study Jupiter across a distance of over 450,000,000 miles. Measurements and photographs obtained from close range will be of absorbing interest, and the dispatch of the first Jupiter probe is eagerly awaited. It should not now be long delayed, and we may hope to secure detailed photographs of the Giant Planet with its belts, its satellites, and its enigmatical Great Red Spot.

Chapter Twelve

SATURN

THE OUTERMOST OF the planets known to the ancients was named by them Saturn, after Jupiter's father. It is not nearly so brilliant as Jupiter, and its yellowish hue makes it appear somewhat leaden; also it moves relatively slowly against the starry background, and our ancestors regarded it as baleful. Yet when seen through a telescope, it is certainly the most beautiful object in the whole sky (Plate XIV).

Everyone knows that Saturn is the planet with the rings, and it is these rings which make it unique. Jupiter is larger and more important in the Solar System as a whole, and both Venus and Mars are far more imposing when seen with the naked eye, but in its own way Saturn is unrivalled.

The splendour of the ring system tends to divert attention from the globe itself, and it is true that surface details are not really prominent. Basically, Saturn is not unlike Jupiter, and it too has its cloud belts and its spots, but there seems to be less activity. Saturn is a more quiescent type of world.

Saturn is appreciably smaller than Jupiter—the equatorial diameter is 75,100 miles, the polar diameter 67,200—and is more remote (Fig. 58). The average distance from the Sun is

Fig 58. Saturn and Earth compared.

148

886,000,000 miles, so that it is always at least 740,000,000 miles away from us. The orbital velocity is 6 miles per second, and the sidereal period is $29\frac{1}{2}$ years. As the spin is rapid, about $10\frac{1}{4}$ hours, a Saturnian 'year' will contain some 25,000 'days'. Neither would it be practicable to divide the year into lunar months, as we do. Our Earth has only one Moon, but Saturn has at least ten.

In size, Saturn is inferior only to Jupiter. Its volume is over 700 times as great as that of the Earth, but its mass is only 95 times as great, because the density is very low by planetary standards—in fact it is less than that of water. Though the escape velocity is high (22 miles per second) the surface gravity is not. Surface gravity depends not only on the mass of a body, but also upon its diameter; for two globes of equal mass, the smaller—and therefore denser—will have the stronger surface pull, because an observer standing there will be closer to the centre of the globe. Because Saturn is so large, a man who weighs 14 stone here would weigh no more than 16 stone there. Apart from Jupiter, there is no planet in the Solar System upon which an Earthman would feel uncomfortably heavy.

Needless to say, this example is theoretical only. It would be rather difficult to stand upon the surface of a gaseous globe whose outer parts are less dense than water; and we may be quite sure that whatever may happen in the future, nobody will ever go to Saturn. Close-range television pictures will be obtained, and eventual bases may be set up on some of the satellites in the Saturnian system. But visiting a giant planet is out of the question, intriguing though the idea may sound!

It seems likely that Saturn and Jupiter are built upon the same pattern. According to Wildt's model, the solid core of Saturn is 28,000 miles in diameter, with an ice shell 6,000 miles deep. This leaves around 16,000 miles for the depth of the atmosphere. On Ramsey's theory the planet is made up of about 60 per cent by mass of hydrogen; the core of metallic hydrogen has a diameter of 25,300 miles, with an 8,000-mile thick layer of more normal hydrogen. The pressure near the core would be about 5 million atmospheres, and the density of the metallic hydrogen there would be about twice that of water. Some recent Russian calculations increase the amount of overall hydrogen in Saturn to as much as 80 per cent by mass.

149

As Saturn is so much more remote than Jupiter, we would expect it to be colder, and the measured temperature is indeed very low; the upper clouds are at about −240 degrees Fahrenheit. Consequently more of the ammonia has been frozen out of the atmosphere, and spectroscopes record a greater amount of methane, which does not freeze so easily. There is no reason to doubt that, as with Jupiter, the bulk of the atmosphere is made up of hydrogen, together with helium and smaller quantities of other constituents. Gallet's model for the outer layers of Jupiter could equally well apply, with modifications, to Saturn; but we must reserve final judgment until we can draw upon the information sent back by passing space-probes.

Telescopes of fair size are needed to show much on the disk, but in general there seems to be a marked resemblance to Jupiter in one of its less active moods. Saturn's belts appear curved; the equatorial zone is usually brightish cream in colour, and once again we have the phenomenon of different parts of the planet rotating at different speeds. The equatorial period is about 10 hours 14 minutes, but in high latitudes this may be increased by twenty minutes or more. Exact information is rather difficult to obtain.

Periodical outbursts of activity take place, particularly near the equator. They may be compared with a very mild Jovian outbreak, and there are times when a moderate telescope can show considerable detail. However, there are no semi-permanent features apart from the belts, and there is nothing comparable with the Great Red Spot on Jupiter.

Major outbursts are rare, and the only really striking one of recent years took place in 1933. In August of that year, the British amateur W. T. Hay (Will Hay, the stage and screen comedian) discovered a prominent white spot in the equatorial zone. It quickly became very prominent, and as a boy of nine I well remember seeing it with my 3-inch refractor. It gradually lengthened, and the portion of the disk following it darkened; subsequently the forward edge of the spot became diffuse in outline, while the following edge remained clear-cut. Sir Harold Spencer Jones, the Astronomer Royal, said that the appearance suggested 'a mass of matter thrown up from an eruption below the visible surface, encountering a current travelling with greater speed than the erupted matter, which was carried forward with the current while still being fed from the following end'. The spot did not last for long. It soon faded, and in a few months it had disappeared completely. White spots have been seen since, but all have been much inferior to Hay's in size and brilliance.

The precise cause of the outbreaks is not known, but the spots are almost certainly of the same nature as those on Jupiter. Saturn's relative quiescence may well be due to its lower temperature and density. Colours on the disk have been reported

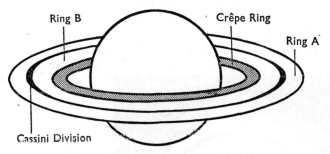

Fig. 59. The ring system of Saturn.

from time to time—greens, browns, and even reds—but personally I have never seen anything of the sort.

It is the ring-system which makes Saturn unique in its glory. A small telescope will show it, though the more delicate features require considerable power. The rings have been known ever since the early seventeenth century, and Galileo observed them, though not clearly enough for him to realize what they were. He mistook Saturn for a triple planet, and was puzzled to find that after a few years the strange aspect vanished, leaving Saturn normal in shape. We have found out the answer to the problem, but Galileo never did.

In 1659 Christiaan Huygens, probably the best observer of his time, issued a famous anagram* in which he announced that Saturn was surrounded by 'a flat ring, which nowhere touches the body of the planet, and is inclined to the ecliptic'.

* The anagram read: aaaaaaa cccc d eeeee g h iiiiiiii llll mm nnnnnnnnn oooo pp q rr s ttttt uuuuu. Rearrange these letters in their proper order, and you will obtain the Latin sentence; Annulo cingitur, tenui, plano, nusquam cohærente, ad eclipticam inclinato. In those days it was quite common for discoveries to be announced in anagram form, to establish priority. Galileo himself had done so when he first detected the phases of Venus.

He was correct so far as he went; but his theory met with a surprising amount of opposition. Honoré Fabri, a French Jesuit mathematician, attacked Huygens bitterly, and claimed that the odd aspect of Saturn was due to the presence of four satellites—two dark and close-in, the other two bright and farther out. Not for some years was Huygens' 'ring' accepted by all astronomers. Some of the drawings made around this time look very peculiar, but this was only to be expected in view of the telescopes used. Reflectors still lay in the future, and even Huygens had to do his best with a very cumbersome long-focus refractor of small aperture.

There are three principal rings, two bright and one dusky. The whole system is of vast extent (Figs. 59 and 60). The outer-

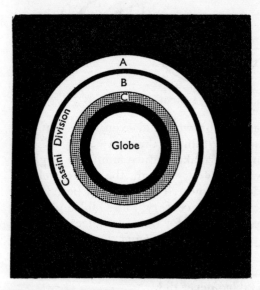

most ring (A) is 10,000 miles wide; then comes a well-marked gap, discovered by G. D. Cassini and hence known as Cassini's Division. It is 1,700 miles wide, and inside it comes Ring B, with a width of 16,000 miles. The ring described by Huygens was a combination of A and B; his telescope was not powerful enough to show the Division, though Cassini discovered it a few years later.

Fig. 60. Bird's-eye view of the ring system of Saturn.

Rings A and B are not alike. B is much the brighter of the two, and is less transparent. Even a small telescope of good quality will show the difference, and the Cassini Division is easy to see when the rings are suitably placed, as in the early 1970s. A 3-inch refractor is adequate.

Inside Ring B, between it and the planet, is a third ring—

Ring C, generally known as the Crêpe or Dusky Ring. It was first recognized in 1850 by two independent observers, W. Bond in America and W. R. Dawes in England. It is by no means striking, and is much less luminous than A or B; also it is transparent enough for the globe of Saturn to be seen through it.

It is strange that the Crêpe Ring should have remained undetected for so long. William Herschel, one of the greatest of all observers, had paid considerable attention to Saturn, using his 49-inch reflector; he observed most of the features we know today, and discovered two of the inner satellites, Mimas and Enceladus, which are very elusive objects. Yet although a few of his drawings show indications of the Crêpe Ring, he did not recognize it for what it is. Nowadays it is not hard to see; in my $12\frac{1}{2}$-inch reflector it is very plain, whereas Mimas and Enceladus have to be looked for carefully. Inevitably the suggestion has been made that the Ring has brightened up since Herschel's time. This I doubt; such an increase in brilliancy would be very hard to explain.

The Crêpe Ring is 10,000 miles wide, and it has always been assumed that between it and the planet is a clear area 9,000 miles in width, into which the Earth would fit comfortably. This has been challenged recently by P. Guèrin, a very experienced French astronomer, using the 24-inch refractor at the Pic du Midi in the Pyrenees. Guèrin claims that his new ring, D, is separated from the Crêpe by a dark division similar to Cassini's; he says, too, that Ring D extends almost down to the globe.

In addition to this, there have been reports of another dusky ring, this time outside Ring A. The original observations were made in France by G. Fournier, in 1907, and have been repeated at various times since. All this is very interesting indeed, but it cannot be said that either Guèrin's Ring D or Fournier's outer ring (also, rather confusingly, referred to as D) can be regarded as established. I have looked for Fournier's ring with very large telescopes, including the 33-inch refractor at Meudon; I have had no success, and if the ring had been as prominent as some observers suggest, I think I would have seen it. On the other hand, I would on no account be dogmatic about it. Overlooking things is only too easy, as I know to my cost!

The shadows cast by the rings on the disk of Saturn are easy to observe, and unwary observers have often mistaken them for belts. The disk can, too, cast shadow on the rings, and recently Gilbert Satterthwaite, Director of the Saturn Section of the British Astronomical Association, has suggested that the profile of this shadow may be helpful in detecting any irregularities in the thickness of the rings.

The ring-system is circular; it seems elliptical to us because we always see it at an angle. The overall diameter of the system is 169,000 miles. Yet the rings are remarkably thin; their thickness cannot be more than about 10 miles, and may be even less. If we reduce Saturn in scale to a globe with an equatorial diameter of five inches, the ring-span will be one foot, but the ring thickness will be a mere 1/1500 of an inch.

This has interesting consequences from an observational point of view. When the rings are edge-on to us, as happened in 1950 and again in 1966—and will happen again in 1980—they almost disappear. To be accurate, they should become virtually unobservable, both when the Earth goes through the ring-plane and when the Sun does so, since in the latter case only the extreme edge of the excessively thin ring can catch the sunlight. It is often said that the rings can vanish completely, even in large telescopes. With this I cannot agree, since in 1966, using the 10-inch refractor at Armagh Observatory, I was able to keep them in view right through the period when the Earth passed the ring-plane—though admittedly they were very faint and difficult. For some time, to either side of the actual passage, the rings appear as nothing more than a thin line of light, and no details in them can be seen.

The edgewise presentation occurs at alternate intervals of 13 years 9 months and 15 years 9 months. The inequality is caused by the eccentricity of Saturn's orbit. During the 13¾-year interval, the south pole of Saturn is tilted sunward, so that part of the northern hemisphere is hidden by the rings; Saturn passes through perihelion, and so is moving at its fastest. During the 15¾-year interval, the north pole is turned sunward, so that parts of the southern hemisphere are covered up; Saturn passes through aphelion, and so is moving more slowly. At the moment (1971) Saturn is excellently placed for observers in Europe and the United States. It is well north of the celestial equator, and the rings are wide open.

The rings lie exactly in the plane of Saturn's equator, which is tilted to the Earth's orbit at an angle of 28 degrees. The diagram in Fig. 61 shows the appearance of the planet for the 1966–1987 period. After the next edgewise presentation, it will be the turn of the southern hemisphere to be covered by the rings as seen from Earth.

The two brightest rings, A and B, look so solid that it was natural for the early telescopic observers to class them as solid sheets. Unfortunately for this theory, the British mathematician Clerk Maxwell showed in 1859 that no such ring could manage to survive, since the whole system lies within the danger-zone of Saturn known technically as Roche's Limit.

In 1848 Édouard Roche, of France, had calculated that if a

Fig. 61. Changing aspects of Saturn's rings.

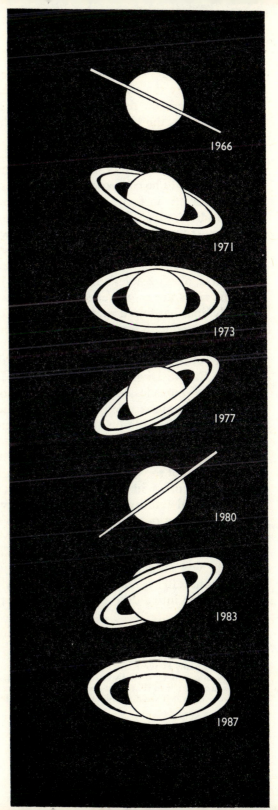

solid body comes within a certain distance of a planet, it will be broken up by the planet's gravitational pull. The limiting distance depends on the size and mass of the planet concerned. With Saturn, the rings are well inside the Limit, and cannot therefore be solid or liquid. If they were, then they would be promptly and unceremoniously torn to pieces.*

J. J. Cassini, son of G. D., made the shrewd suggestion that the rings might be composed of numerous small solid particles— tiny moonlets, in fact—each revolving round the planet in its own individual orbit. This has turned out to be the correct answer. It fits all the facts, and explains why the inner parts of the ring-system move round Saturn more rapidly than the outer. Each ring-particle behaves as though independent, and, of course, obeys Kepler's Laws.

The rings are more reflective than the globe of the planet, and Kuiper has found that they must be either made up of ice or else coated with some icy substance; ammonia ice, not water ice, is a strong probability. We cannot do more than guess as to the size of the particles, but they can hardly be very big.

The Cassini Division, between Rings A and B, is a genuine gap. Ring particles tend to avoid it, and it is suggested that Saturn's satellites are responsible—particularly Mimas, Enceladus and Tethys. (Janus, the closest-in member of the family was only found recently, and as yet its movements are not well known.) Mimas moves only 30,000 miles beyond the outer edge of Ring A, not so very far outside Roche's Limit, and it may be up to 300 miles in diameter, so that it has an appreciable gravitational pull even though its density is low. It has been found that any particle moving in Cassini's Division must have a revolution period which is half that of Mimas, one-third that of Enceladus and one-quarter that of Tethys. Therefore the particle will suffer regular perturbations, which will be cumulative; and after a relatively short time it will be driven out of the Division into a different orbit. In effect, the Division is kept 'swept clear'.

Various other divisions in the ring-system have been reported;

* The Roche figures refer only to a body which has no gravitational cohesion. Otherwise, many of the Earth's artificial satellites would be disrupted, since they are very close to the ground by astronomical standards. All the same, simple calculations are enough to show that any solid or liquid ring for Saturn is out of the question—even if it were plausible in other ways, which it is emphatically not.

one list that I have seen gives as many as ten. Yet, of these, only one is well-authenticated. This was discovered by the last-century German astronomer, Johann Encke, and is named after him. It lies in Ring A, and I am disposed to accept its reality, for the excellent reason that I have often seen it. However, it is visible only at the ends of the rings (the ansæ), and is always elusive. It may not be a division in the same sense as Cassini's, and the other gaps which have been listed may be nothing more than surface 'ripples' in the ring—if, indeed, they exist at all.

The origin of the ring system makes up an interesting problem. The rings may be due to the break-up of a former satellite which came too close to Saturn, but it is equally likely that the ring-particles have never formed one body, in which case they presumably come from a cloud of material which used to surround Saturn.

Of the ten satellites, the most important is Titan, which is of planetary size, and has an escape velocity of about 1·7 miles per second. Its diameter is rather uncertain; estimates range between 3,500 miles and only 2,700 miles, but either way it is comparable with Mercury, and is a good deal larger than our Moon. It moves round Saturn in an almost circular orbit at a distance of over 750,000 miles; the period is about 16 days. Surface details have been made out with giant telescopes, but are never easy.

Titan has the distinction of being the only satellite in the Solar System known to have an atmosphere. The discovery was made in 1944 by G. P. Kuiper, using spectroscopic methods. The atmosphere is thin, and apparently made up of methane, so that Titan is hardly a welcoming world—particularly in view of its low temperature.

It may be asked why Titan (escape velocity 1·7 miles per second) has an atmosphere, while Mercury (2·6 miles per second) has not. The reason is that Titan is colder, so that the particles round it are less agitated, and move less quickly. Even so, it is a borderline case. If the temperature were raised by as little as 100 degrees Fahrenheit, it has been calculated that the atmosphere would escape.

Titan was discovered by Huygens in 1655, and is an easy object with a small telescope, but the remaining members of

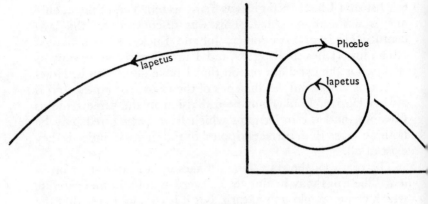

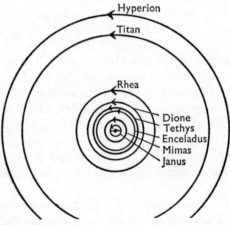

Fig. 62. Orbits of the satellites of Saturn. *(Upper inset.)* The orbit of Phœbe, much the most remote satellite, with that of Iapetus for comparison; this inset is of course drawn to a different scale.

Saturn's family are much fainter. Their diameters are not known with accuracy, since all we can really do is to assume average reflecting powers and then calculate according to the apparent magnitudes.

Starting from the inner part of the system, we come first to Janus, which was discovered by Audouin Dollfus, at the Pic du Midi, in December 1966 (Fig. 62). This was the year of the edgewise presentation of the rings, so that the closer satellites were particularly well placed for observation. Janus is highly elusive; it cannot be more than 200 miles across, and it is impossible to observe except when the rings are virtually out of view. As yet we know very little about it.

I cannot resist telling a story against myself. During 1966 I had been making a series of observations of Saturn with the 10-inch refractor at Armagh, in Northern Ireland (Fig. 63). I had made estimates of the known satellites, including Mimas and Enceladus; but having no tables to hand I had not reduced the observations systematically. I was waiting until the series had been completed, which would have been during the following January. After Dollfus' announcement, I did some checking, and found that I had recorded Janus several times between July and November—quite without realizing that I

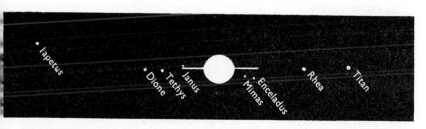

Fig. 63. Saturn's satellites, October 24, 1966 (21.10 hours), drawn by Patrick Moore with ×400 on the Armagh 10-in. refractor. Janus was recorded, but was not recognized as new!

was looking at a new satellite! Of course, I can claim absolutely no credit; it was a good example of overlooking the unexpected. The sole value of my observations was that I was able to show that Janus is slightly brighter than was at first thought. Alas, nobody will see it again until 1980.

Next come Mimas and Enceladus, both found by Herschel in 1789 with his newly-completed reflector; the mirror was 49 inches across, and the focal length was 40 feet. Both Mimas and Enceladus are larger than Janus, but are very insubstantial; they may be icy in nature, and perhaps even less dense than water. Tethys is larger, but of the same type. Dione, however, is different. It is the densest of all Saturn's satellites, with a specific gravity of 3·2 (about the same as that of our Moon), and it is much more massive than the inner four. Official estimates make Dione slightly fainter than Tethys, but my own observations, made between 1950 and the present time, make it a little brighter on average—though both are variable to some extent.

Rhea is larger still, and may have a diameter of 1,000 miles; it is not difficult to see with a 3-inch refractor. Beyond it comes Titan, and then another small body named Hyperion. Because of its faintness, Hyperion is not an easy object, and is best located when close to Titan in the sky.

Iapetus, discovered by G. D. Cassini in 1671, is probably the most interesting member of the satellite family. It varies in brightness; when west of Saturn it is much more conspicuous than when to the east. It has a revolution period of 79 days, and is more than two million miles from Saturn, so that it is not a rapid mover.

Here again the estimates of magnitude are not in good accord, and I propose to have the temerity to give my own results.* I find that at its best, Iapetus is much brighter than Rhea, and not far inferior to Titan, though when at its faintest it is dimmer than either Tethys or Dione. If I am right, then Iapetus must be larger than the official values given. It could, I suggest, be 1,500 to 1,800 miles in diameter, though in some lists it is given as only 700 miles.

Obviously, some peculiarity of the satellite itself is responsible for its fluctuations. It may be that Iapetus has a synchronous or captured rotation, so that it spins once in the same time that it takes to go round Saturn (79 days), in which case its variations could be explained by an unequal reflecting surface. F. L. Whipple has suggested that in the remote past Iapetus was either disfigured by a collision with a wandering body, or else discoloured by gaseous outbursts from Saturn—but on the latter theory why should Iapetus, and Iapetus alone, have been affected? Moreover, the case of Mercury has made us less confident about synchronous rotations than we used to be.

If Iapetus had an atmosphere, we might explain the variations by the freezing of the gases during the long night, so that the brightness at western elongation would be due to the

* Since I published my first paper on the subject, in 1969, my estimates have been largely confirmed by the American observer K. Delano.

Sun's rays striking this frozen, more highly reflective surface deposit. Unfortunately we would still have to assume a captured rotation, and in any case it seems most improbable that Iapetus, with its low escape velocity, can retain any atmosphere at all. On the whole, it looks as though we must come back to the idea either of a satellite which is irregular in shape, or else a body which has one hemisphere much more reflective than the other. At any rate Iapetus is well worth watching, if only because we cannot yet be sure that the variations do not show slow, long-period changes relative to the position in orbit.

The outermost of Saturn's satellites, discovered photographically by W. H. Pickering in 1898, is named Phœbe. Although very small, with an estimated diameter of only 150 miles, it has an interesting orbit, fairly circular but highly inclined; it is so far from Saturn—over eight million miles—that it takes a year and a half to complete one revolution. Like the outermost members of the Jovian family, it moves in a retrograde or wrong-way direction, and like them it may be a captured asteroid rather than a genuine satellite.

Eclipses, transits and shadow-transits of the satellites may be observed, though less easily than with the Galilean moons of Jupiter. Only for Titan are the phenomena within the range of a small telescope, and this is a pity, because the orbits of the smaller satellites are not known with absolute precision, and timings of the eclipses and transits would be useful—particularly in the case of the rare eclipses of Iapetus. Mutual phenomena have been seen now and then; for instance, on April 8, 1921 Major A. E. Levin and L. J. Comrie observed an eclipse of Rhea by the shadow of Titan.

Pickering's discovery of Phœbe was made with a 24-inch telescope at Arequipa in Peru, the southern station of Harvard College Observatory. Six years later he announced that he had found a new satellite, with an orbit between those of Titan and Hyperion. He named it Themis, and for some time its existence was regarded as well-established; but it has never been reported since, and probably does not exist.

Until recently it seemed that Saturn must be regarded as well beyond the range of space-probes, manned or unmanned. Certainly we cannot yet consider sending astronauts out to these remote parts of the Solar System; but there is every reason to suppose that an automatic vehicle will pass by Saturn before the end of the 1970s. Plans for such a probe are already well advanced, and there is no real reason to think that anything will delay them. When the first probe nears the Ringed Planet, and sends back its data, we shall learn more than we could ever do otherwise. We may find out whether Saturn, like Jupiter, has a strong magnetic field; we should obtain spectacular photographs of it from close range, and we should be able to decide whether either of the two dubious rings really exists. We may even solve the minor but still interesting problem of Iapetus.

Though the rings are so magnificent, they have very little mass, and are not actually so important as they look. Theoretically the globe is much the more significant. After all, Saturn is much the largest and most massive of the planets, except for Jupiter. It too seems to send out more energy than it should do, even though there is no chance that the outer layer is hot. On the other hand, it is not a strong, erratic radio source of the same kind as Jupiter.

From time to time there have been suggestions that both Jupiter and Saturn may support some very low, unfamiliar form of life. It is true that there may be plenty of ice and water-vapour below the outer clouds, and there must be a region in which the temperature is tolerable by our standards; but any life would have to be entirely different in nature from our own, and as yet there is no evidence in favour of anything of the sort. I am willing to be proved wrong; but I am ready to state my personal belief that there is no living material on Jupiter, Saturn, or the other outer planets.

One day, no doubt, men will stand upon some of Saturn's satellites and admire the view. The most spectacular pictures of the planet itself will be obtained from Phœbe and Iapetus, since only these two satellites have orbits which are appreciably inclined to the plane of the rings; the others move practically in the ring-plane, which also coincides with that of Saturn's equator. Go to Dione, for instance, and the rings will be permanently edge-on. Yet even so, the sight will be magnificent, and it is a pity that we of the twentieth century will never see it. Meanwhile, we can at least enjoy the spectacle of Saturn

from our home on Earth. As the planet swims into the telescopic field—lonely, remote and unutterably magnificent—it
is a sight which nobody who has seen it will ever be likely to
forget.

Chapter Thirteen

URANUS

TO THE ANCIENTS, Saturn marked the boundary of the planetary system; beyond it, at a much greater distance from us, came the sphere of the fixed stars. Nobody gave any serious consideration to the possibility that there might be another planet beyond Saturn. In any case, the Sun, the Moon and the five planets already known made a grand total of seven, and seven was the 'mystical' number. Nothing could be tidier or more predictable.

Kepler's suggestion of an extra planet between Mars and Jupiter was not really relevant, since any such body would certainly be small. There matters stood until 1781, when a discovery by a then unknown musician-astronomer took the scientific world by surprise.

William Herschel was born in Hanover, but came to England when still a young man, and became a professional organist. His hobby was astronomy, and he began to make reflecting telescopes. Gradually he became more and more skilful, and decided to undertake an ambitious 'review of the heavens', to be carried out with reflectors of his own construction. It was a long-term project; Herschel was well aware of the labour involved, but he had great confidence in his own ability.

On the night of March 13, 1781 he was using a 7-inch reflector to examine stars in the constellation of Gemini, the Twins, when he made the observation which was to alter his whole life. To quote from his own account:

'In examining the small stars in the neighbourhood of H Geminorum I perceived one that appeared visibly larger than the rest; being struck with its uncommon appearance I compared it to H Geminorum and the small star in the quartile between Auriga and Gemini, and finding it so much larger than either of them, I suspected it to be a comet.'

Comets, though interesting, are by no means uncommon, and the discovery did not cause Herschel any particular excitement at the time. His paper on the subject was headed 'An Account of a Comet', and he had no idea of the importance of the observation. Then the mathematicians set to work, and the orbit of the body was worked out. It was not in the least like that of a comet. Instead, the object was a planet, much more remote than Saturn, and moving round the Sun in a period of 84 years at a distance of 1,782 million miles.

At first the planet was known as the Georgium Sidus, or Georgian Star, in honour of King George III of England and Hanover, who raised Herschel to the status of King's Astronomer (not Astronomer Royal), and gave him a yearly grant which enabled him to give up music as a profession and devote all his time to astronomy. Foreign astronomers were not impressed, and neither did they accept the alternative name of Herschel, in honour of the discoverer.* Finally the mythological system was followed, and the new planet became Uranus, after the original ruler of Olympus.

It is often said that the discovery was due to sheer chance, but this is unfair to Herschel. As he said in a letter written to a friend of his, Dr. Hutton: 'Had business prevented me that evening, I must have found it the next, and the goodness of my telescope was such that I must have perceived its visible planetary disk as soon as I looked at it.' Incidentally, the new planet could hardly have been missed for much longer even if Herschel had never made a telescope. It would certainly have been found during the search organized by Schröter and von Zach in the first years of the nineteenth century, when they set out to hunt for the missing planet between Mars and Jupiter.†

Though Herschel was the first to recognize Uranus, he was not the first to see it. It had been recorded on several previous occasions—notably by the first Astronomer Royal, John Flamsteed, who saw it six times between 1690 and 1715 without

* The well-known amateur Admiral Smyth, much later, made a telling point. The name 'Herschel' might be tolerable, as would 'Le Verrier' for the co-discoverer of Neptune in 1846; but suppose either planet had been discovered by contemporary astronomers who rejoiced in the names of Bugge and Funk?

† In 1955 I had a letter from a boy who had been using a home-made telescope constructed out of spectacle-lenses and cardboard tubes. He had been looking at star-fields, and wrote to me to ask about a curious body which was not in his maps. It was, of course, Uranus. The episode was interesting, because it showed that the planet is non-stellar in aspect even to someone with very modest equipment and with no knowledge of its position.

realizing that it was anything but an ordinary star. Keen-sighted persons can see it with the naked eye, provided that they know where to look; its stellar magnitude is officially given as 5·7, though many estimates make it rather brighter than this.

Uranus is a giant planet. Though it is much smaller than Saturn or Jupiter, it is far larger than the Earth (Fig. 64), and its equatorial diameter has been measured as 29,3000 miles. The polar diameter is rather less, since the polar flattening is about as marked as with Jupiter. Uranus is denser than water,

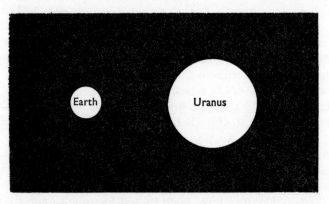

Fig. 64. Uranus and Earth compared.

and much denser than Saturn. Yet although its volume is roughly 50 times that of the Earth, its mass is only 14½ times as great; the escape velocity is 13·9 miles per second, but the surface gravity is only fractionally greater than that of the Earth. If it were possible to stand on Uranus, a man who weighs 13 stone at home would find his weight increased by only one stone.

Of course, it would be quite impossible to stand on Uranus, because here again we are dealing with a world whose outer layers are made up of gas. Methane is very much in evidence spectroscopically, but there are few signs of ammonia, because Uranus is so cold (around —310 degrees Fahrenheit at the top of the cloud layer) that almost all the ammonia has been frozen out. Hydrogen and helium are certainly plentiful.

Although Uranus is much more similar to Jupiter or Saturn than to the Earth, there are marked differences. On Wildt's original pattern (Fig. 65) there was assumed to be a rocky core 14,000 miles in diameter, with a 6,000-mile ice-layer and a gaseous atmosphere 3,000 miles deep; it was thought that the density at the core would be about that of ordinary rock. The Wildt pattern for Uranus was, in fact, essentially the

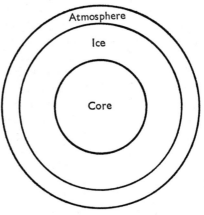

Fig. 65. Structure of Uranus, according to Wildt's theory.

same as for Jupiter and Saturn, but on later theories the differences are more noticeable.

Ramsey, for instance, believed that because Uranus is so much less massive than Jupiter or Saturn it has lost much of its original hydrogen and helium; since these are the two lightest gases, they are the most prone to escape unless held down by a very strong gravitational pull. Ramsey's pattern described a planet made up largely of ammonia, methane and water, in which case Uranus would really be intermediate in type between a 'Jovian' planet and a 'terrestrial' one.

In 1970 some new calculations were made by two Russian astronomers, V. Zharkov and V. Trubitsyn of Moscow, who concluded that although the cloud-layer of Uranus is extremely cold, the core is likely to be hot; perhaps as much as 30,000 degrees Fahrenheit. This would mean a temperature difference of several thousands of degrees between the 'core' and the outer layer, assuming of course that a definite boundary exists. As yet calculations of this sort are very difficult to check; space-probes may help when they can be sent out to Uranus, and we shall also find out whether or not Uranus has as strong a magnetic field as Zharkov and Trubitsyn expect. No radio emissions have so far been detected; but across so vast a distance, even powerful planetary radio waves would be hard to identify.

Uranus, like Jupiter and Saturn, is a quick spinner. The best value for the rotation period so far derived is 10·8 hours, in which case there are about 65,000 'days' in each Uranian year. Probably the rotation is quickest in the equatorial zone, as with Jupiter, but we have no positive information.

The strangest fact about Uranus is the tilt of its axis. As we have noted, most of the planets rotate with their axes not sharply inclined to the perpendiculars to their orbits; the Earth, Mars, Saturn and Neptune have inclinations of between 23 and 30 degrees, while that of Jupiter is only just over 3 degrees. Uranus has its own way of behaving. The axial tilt is 98 degrees, which is more than a right angle, so that the rotation is technically retrograde (Fig. 66).

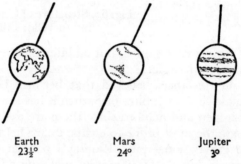

Earth
23½°

Mars
24°

Jupiter
3°

Fig. 66. Axial Inclination

This means that the seasons on Uranus are peculiar, to put it mildly. First much of the northern hemisphere, then much of the southern, will be plunged into Stygian blackness for twenty-one Earth years at a time, with a corresponding midnight sun in the opposite hemisphere. For the rest of the revolution period, 48 Earth-years, conditions will be less extreme.

The unusual tilt of the axis means that from Earth we sometimes look straight at a pole; sometimes at the equator, as shown in Fig. 67. In 1946 the south pole of Uranus lay in the centre of the disk. By 1966 the equator ran straight 'up and down', with the poles lying on the limbs; by 1985 we will have a bird's-eye view of the north pole, and by 2007 the equator will again be presented.

Because Uranus is markedly flattened, the changing presentation will have an effect upon the apparent brightness. (This can be shown by the simple experiment of holding up an egg first with the thin end presented and then broadside-on. The surface area will be greater, to the observer, in the latter case—though naturally the relative flattening of Uranus is much less than with the egg.) Other variations can be expected from the eccentricity of the orbit, since Uranus will be most brightly illuminated when near perihelion, and will also be closer to the Earth. Yet even when all these factors are taken into account, it seems that the magnitude changes are not fully explained. In 1917 W. W. Campbell, at the Lick Observatory, found a small regular change which he associated with the 10·8-hour

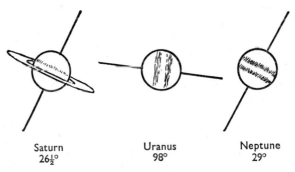

Saturn	Uranus	Neptune
26½°	98°	29°

of the planets.

rotation period; Joseph Ashbrook, between 1935 and 1948, confirmed an earlier suggestion that there seemed to be an 8·4-year variation with an amplitude of one-third of a magnitude. On the other hand, some observers are inclined to discount these results, and to consider that Uranus is to all intents and purposes steady in its light.

The question is more important than might be thought, because any variations not directly linked with distance or rotation would presumably be due to activity on the top of the cloud-layer of Uranus. Unfortunately, surface details are very difficult to see even with powerful telescopes. There seems to be a brightish equatorial zone, with faint belts to either side; occasional white spots have been reported, but the amateur with modest equipment cannot hope to see anything on the dim, decidedly greenish disk.

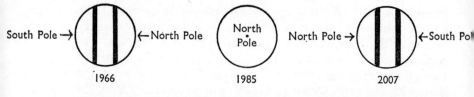

Fig. 67. The changing presentation of Uranus. In 1966 the equator appeared crossing the disk; in 1985 the north pole of the planet will appear central; by 2007 the equator will be presented again.

The most valuable project for the amateur, therefore, is to make estimates of the brightness of the planet. Visual observations are accurate enough to be useful, and can be carried out in the same way as for variable stars.* Oddly enough, binoculars are ideally suited to this sort of investigation. Uranus is too faint to be easily seen with the naked eye, and a telescope shows it as a noticeable disk, hard to compare with the points of light which are the stars. I found this also in May 1955, when Uranus lay close to Jupiter in the sky. From May 8 to 11 the two planets were in the same telescopic field, and I decided to estimate the magnitude of Uranus by comparing it with the four Galilean satellites of Jupiter, using various magnifications on my 12½-in. reflector. Alas, I failed. Uranus appeared larger than any of the satellites, but it was also much dimmer, area for area; and I found myself quite unable to make any accurate comparisons.

It is hardly necessary to add that Uranus and Jupiter were not really close together. They simply happened to lie in the same direction as seen from Earth. Uranus is more than three times as far away from us as Jupiter.

There must be occasions when Uranus passes in front of a star, and occults it. Timings would be useful, both in giving the exact position of the planet and in leading to measurements of its apparent diameter, but as yet (1971) no such occultation has ever been observed. The next opportunity will occur in 1977, when Uranus is due to occult a star of magnitude 8·8.

* See the relevant sections of my book *The Amateur Astronomer* (7th edition; Lutterworth Press, 1971.)

As befits its status as a senior member of the Solar System, Uranus is accompanied by a retinue of satellites (Fig. 68). Herschel believed that he had discovered six, but for once he was mistaken; four of his 'satellites' turned out to be faint stars, and only two, those now known as Titania and Oberon, were genuine. Two more, Ariel and Umbriel, were found

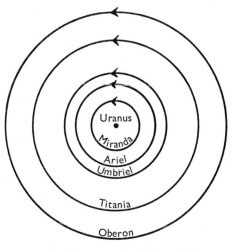

Fig. 68. Orbits of the satellites of Uranus.

in 1851 by the English amateur William Lassell, while the fifth and faintest, Miranda, was detected by G. P. Kuiper at the McDonald Observatory, Texas, in 1948.

All five satellites revolve almost in the plane of Uranus' equator; thus their orbits appear circular when either of the planet's poles is presented to us (as in 1946 and 1985), almost linear when the equator is displayed (1966, 2007)—see Fig. 69. Their distances from Uranus range between 76,000 miles for Miranda out to 364,000 miles for Oberon.

The magnitudes of the four main satellites have often been underestimated. Titania is usually the easiest to see, though some careful measurements made in 1947 by W. H. Steavenson, using his 30-inch reflector, indicated that Ariel is actually the brightest of the four; it is, however, much closer to the planet. Steavenson also found that both Titania and Oberon vary considerably, though it would be premature to suggest that there is any similarity between their behaviour and that of Iapetus. Satellite phenomena are excessively difficult to observe. So far as I know, the only record of a shadow transit was obtained by Cragg and Wilson, using the 100-inch reflector at Mount Wilson in California, some years ago.

Adopting Steavenson's revised estimates of brilliancy, and assuming that the satellites are of normal reflecting power, it

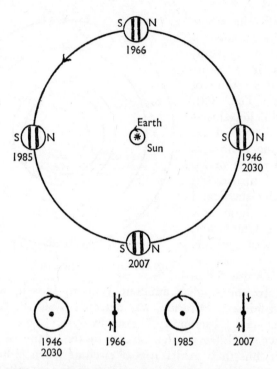

Fig. 69. Changing aspects of Uranus. *(Upper)*
Pole and equator. *(Below)* The apparent
orbits of the satellites.

seems that Ariel, Titania and Oberon are between 1,800 and
1,300 miles in diameter, Umbriel between 700 and 900, and
Miranda 200; but these values are most unreliable. At any
rate, the escape velocities must be low, so that none of the five
can have retained any trace of atmosphere.

Uranus is hopelessly out of range for manned rockets of the
present day, but, as we have noted, the Grand Tour technique
should result in an automatic probe by-passing the planet in
1985. The mission will be designed to obtain information about
Uranus' mass, temperature, composition, magnetic field and
other characteristics. It is also hoped to receive television pictures as good as those sent back from Mars by Mariner 4 in
1965.

Looking further ahead, perhaps for several centuries, we can

picture the time when manned flights to Uranus become possible. From there—from one of the satellites, say—the Sun will
have an apparent diameter of only $1\frac{1}{2}$ minutes of arc, or less
than twice the apparent diameter of Jupiter as seen from Earth;
but it will still be extremely brilliant, and will cast as much
light as more than a thousand full moons. Little could be seen
of the other planets. Saturn would be a naked-eye object, but
would remain close to the Sun in the Uranian sky, never moving
much further away than Mercury does to us; it would be an
inferior planet, best seen every $22\frac{3}{4}$ terrestrial years. Jupiter
would never be more than about 17 degrees from the Sun, and
would seldom be visible without optical aid. Neptune, of
course, would be reasonably bright when at or near opposition,
but it would be out of view for long periods when it and Uranus
are on opposite sides of the Sun. Remember, too, that on average Uranus is only one and a half times closer to Neptune than
we are. Maps tend to be deceptive, and it is quite wrong to
suppose that the two worlds are near neighbours—just as some
Europeans have the mistaken idea that New Zealand is almost
within hailing distance of Australia!

There should be no difficulty in locating Uranus. During the
period between 1970 and 1977 it remains in Virgo; it then
enters Libra, and not until 1982 will it reach Scorpio. Binoculars will show that it is unlike a star, and telescopes will magnify
it into a distinct greenish disk. It is an interesting world, and it
has the distinction of being the first planet to be discovered by
modern man.

Chapter Fourteen

NEPTUNE

FAR AWAY IN the depths of space, a thousand million miles beyond Uranus, lies the last of the giant planets—Neptune, a world so remote that we cannot see it at all without optical aid. Neptunian astronomers, if they existed, could know nothing about the Earth; but strangely enough, Earth astronomers knew about Neptune before they actually observed it. The story of its discovery is certainly worth re-telling.

The key to the problem was provided by Uranus. As we have noted, Uranus had been recorded several times before Herschel found it in 1781; Flamsteed had made half a dozen observations of it, and so had a French astronomer named Le Monnier. Flamsteed did not check these particular observations, and Le Monnier was not blessed with an orderly and methodical mind. However, when the mathematicians came to work out the orbit of the newly-identified Uranus, the old observations came in most useful—even one of Le Monnier's which was pencilled on the back of a bag which had once contained hair perfume!*

Altogether, records of Uranus extended back over a hundred years before 1781, more than one complete revolution of the planet, and it should have been possible to compute a reliable orbit. Unfortunately the old observations did not seem to fit properly with those made after 1781. Something was wrong somewhere, and eventually a French mathematician, Alexis Bouvard, rejected the old observations altogether and worked out a new orbit based only upon positions measured after Uranus had been recognized as a planet.

Even this would not do. Uranus refused to behave; it persistently strayed from its predicted path. Up to 1822 it seemed to move too rapidly, while after 1822 it lagged. It became

---

* Le Monnier seemed fated to miss perpetuating his name. While engaged on cataloguing the stars near the north celestial pole, he introduced an entirely new constellation, Tarandus (the Reindeer) which was promptly forgotten. Le Monnier was undoubtedly a clever scientist, and did much valuable work, but it is said of him that he quarrelled with every person with whom he came in contact.

painfully clear that there must be some unknown factor to be taken into account.

Each planet perturbs the rest; thus the Earth's orbit is appreciably affected by Venus, Mars and even Jupiter. So far as Uranus was concerned, the most powerful disturbing agents were Jupiter and Saturn, but Bouvard had allowed for them both, and still Uranus wandered.

In 1834 the Rev. T. J. Hussey, Rector of Hayes in Kent, made a most interesting suggestion. Suppose that an unknown planet were pulling upon Uranus? This might account for the irregularities in motion, and by 'working backwards', so to speak, it would be possible to track down the planet responsible.

Hussey went so far as to write a letter to George (afterwards Sir George) Airy, the Astronomer Royal of the time, but Airy was not encouraging. There matters rested until 1841, when a young Cambridge undergraduate, John Couch Adams, made up his mind to attack the problem as soon as he had taken his mathematical degree. He passed his final examinations— brilliantly—in 1843, and then began to study the movements of Uranus in earnest. By the end of the year he had worked out just where the new planet ought to be, and, naturally enough, he sent his results to Airy.

Now began a series of misfortunes which led to a most un-dignified dispute in after years. Airy, partly through a lack of confidence in Adams and partly through a misunderstanding, took no action. Delay followed delay, until in 1846 Urbain Le Verrier, a French astronomer, published a memoir which showed that he had approached the problem much as Adams had done—and with similar results.

As soon as Airy saw Le Verrier's memoir, he asked two observers to begin searching in the place indicated by Adams. One was Professor Challis, official astronomer at Cambridge University, and the other a well-known amateur, William Lassell. Still there were delays. Challis had no suitable star-maps of the area, while Lassell was rendered *hors de combat* by a sprained ankle. Challis actually recorded the object he was seeking on August 4 and again on August 12, but he failed to check his observations; and before he had done so Johann Galle and Heinrich d'Arrest, working at Berlin Observatory upon Le Verrier's calculations, had found and identified the new planet.

Adams had been the first to forecast the planet's position; Le Verrier's work had led to the first actual identification, and both mathematicians were worthy of the highest praise. Unhappily, they were made the centre of a childish squabble about priority which is best forgotten, and in which neither of the principals took much part. Wrangles of this sort cannot be defended. It is the discovery which matters, not the man who makes it.

Neptune, as the new planet was named, proved to be a giant of the same type as Uranus. Its average distance from the Sun is 2,793 million miles, and it has a period of 164¾ years, so that not until the year 2011 will it be back at the position in its orbit in which it was first recognized by Galle and d'Arrest. The inclination of its axis is conventional (about 29 degrees) so that the peculiar seasonal effects of Uranus do not occur. The rotation period is rather uncertain. Most authorities give 15·8 hours, but the latest information shows that this may be too long, and that 14 hours is a more probable value.

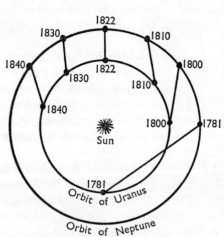

As soon as Neptune was discovered, the orbit of Uranus was recalculated, and this time the old observations of Flamsteed and Le Monnier fitted almost perfectly into place. Neptune was at opposition, so far as Uranus was concerned, in 1822; the Sun, Uranus and Neptune were then almost in line, with Uranus in the mid position. Before 1822, Neptune was tending to pull Uranus along, while after 1822 the effect was reversed (Fig. 70). Had Uranus and Neptune been on opposite

Fig. 70. The pull of Neptune on Uranus. Before 1822 Neptune tended to accelerate Uranus; after 1822, to retard it.

sides of the Sun during the early nineteenth century, the disturbing effects upon Uranus would have been inappreciable, and—apart from sheer chance—the discovery of Neptune would have been considerably delayed.

There is another hint of trouble, however. At the present moment (1971) Neptune is about five seconds of arc away from its theoretical position, which is not very much, but is enough to be disquieting. Pluto, whose orbit lies for the most part outside Neptune's, may be the cause of the discrepancy, because we are still very uncertain about its mass. Alternatively, there may be a still-undiscovered trans-Plutonian planet which is making its presence felt. Either way, the orbit of Neptune remains something of a problem to students of celestial mechanics.

Physically Neptune is very similar to Uranus. It is too faint to be seen with the naked eye; the best modern values make its magnitude 7·95 (slightly below the official 7·7), so that it is an easy object to binoculars, though it looks very starlike unless a telescope of some power is used.

In 1899 E. E. Barnard measured the planet's diameter, and made it between 32,000 and 33,000 miles, rather larger than Uranus. There has never been much doubt that Neptune is the more massive of the two ($17\frac{1}{4}$ times the mass of the Earth, as against $14\frac{1}{2}$ Earths for Uranus), and it was therefore rather surprising when, in 1949 and 1953 respectively, measurements made independently by Kuiper in America and H. Camichel in France reduced the diameter to less than 28,000 miles. In this case Neptune would have a density over twice that of water, much higher than that of Uranus, and the surface gravity would be 1·4 times that on Earth, greater than for any other planet apart from Jupiter.

Neptune's apparent diameter is only a little over two seconds of arc, so that accurate measurements are very hard to make. Luckily there is an alternative method: that of occultations.

As Neptune crawls slowly against the starry background, it must sometimes pass in front of a star. It moves so gradually, and seems so small, that occultations are rare; but if a star is covered, the duration of the occultation will yield a value for the diameter. At Herstmonceux, the modern site of Greenwich Observatory, Gordon E. Taylor made a long investigation, and found that a suitable star would be occulted on April 7, 1968. He organized a programme of observation, and results came in from places as far afield as Australia, New Zealand and Japan. When these results were analysed, the diameter of Neptune was found to be 50,940 kilometres (31,500 miles) as measured through the equator, and 49,920 kilometres (31,100 miles) as measured through the poles. The probable error was a mere 87 miles in each case. So Neptune really is larger than Uranus, and its density comes down to a value of 1·8 times that of water, almost the same as that of Uranus (Fig. 71).

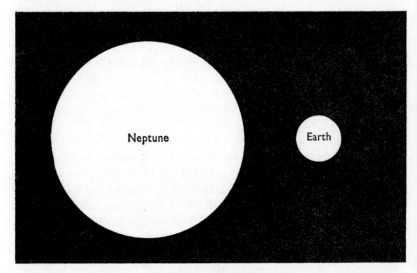

Fig. 71. Neptune and Earth compared.

The temperature of the cloud-layer is lower than with Uranus, because Neptune is farther away from the Sun, but the structures of the two planets are probably similar. Wildt's model gives a 12,000-mile core, over which is an ice layer 6,000 miles thick, while models of the Ramsey type assume that the globe contains a great deal of hydrogen and helium together with vast amounts of water, methane and ammonia. Methane is detectable in the spectrum; ammonia is not, since the very low cloud temperature of —360 degrees Fahrenheit is enough to freeze the ammonia out of the upper atmosphere.

Zharkov and Trubitsyn believe that Neptune, like Uranus, has a very hot central core. Certainly the measured temperatures seem rather less cold than they ought to be, so that the effects of internal heating cannot be ruled out.

Some people describe Neptune as greenish. To me it always appears to have a decidedly blue cast, but the disk is too small to show any detail in moderate-sized telescopes. Even giant instruments can reveal nothing except, possibly, a brightish equatorial zone. Fortunately Neptune is one of the targets of the planned Grand Tour probe of 1977, and by 1988 we ought to have close-range television pictures of it.

Soon after its discovery, some observers—Lassell and Challis in England, and Bond in America—reported indications of a ring, similar to Saturn's but much fainter. This has not been confirmed since, and seems to be definitely absent. But if Lassell was wrong about the ring, he was successful in discovering the brighter of the two satellites, Triton, only three weeks after the identification of Neptune itself.

Triton is one of the largest and most massive satellites in the Solar System. Estimates of its diameter range between 2,300 miles and over 3,000 miles, but its density is believed to be over five times that of water, so that its escape velocity is high enough for it to retain a thin atmosphere. In 1944 Kuiper

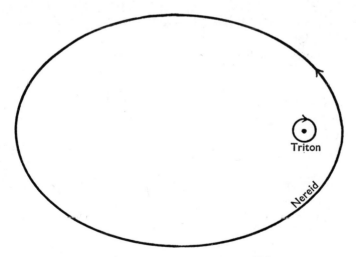

Fig. 72. Orbits of the two satellites of Neptune.

announced that he had found traces of a methane atmosphere, as with Titan. Confirmation is lacking as yet, but there seems no reason to doubt that the atmosphere does exist. A moderate telescope will show Triton, since it is brighter than any of the satellites of Uranus. Its orbit is almost circular, but is sharply inclined; the direction of movement is retrograde. Triton is the only large satellite in the Solar System to have retrograde motion, which may be significant (Fig. 72).

Some calculations made by T. B. McCord indicate that Triton is slowly approaching Neptune, so that eventually it will either collide with Neptune or else be disrupted, perhaps even being spread round in a genuine ring. Yet even if this disaster happens, it lies so far in the future that we cannot speculate about it with any confidence. Moreover, the calculations may be misleading; other mathematicians have cast serious doubts upon them.

The second satellite, Nereid, was discovered by Kuiper in 1949. It is only about 200 miles in diameter, and has an extraordinary orbit, so eccentric that the distance from Neptune ranges between 867,000 miles and over 6,000,000 miles. The period is almost a year, and the movement is direct (west to east). Nereid is quite unlike Triton, and may be a captured asteroid, though admittedly it is strange to picture any asteroid moving in those remote regions. It is so faint that it has never been seen visually, though it has left its image upon photographic plates.

If Uranus is lonely and far-away, Neptune is even more so. If it were possible to go there, little would be seen of the Solar System apart from the shrunken Sun; of all the planets only Uranus and Pluto would be in evidence, and then only when suitably placed. For long intervals they would be out of view. Saturn, Jupiter and the rest would be hard to detect; remember, Neptune is farther away from Saturn than we are.

For the moment we must admit that our knowledge of Neptune is very incomplete. Yet we must not be pessimistic; within twenty years from now we may be able to study television pictures of the bitterly cold gas-clouds which mask the outermost of the giant planets.

Chapter Fifteen

PLUTO

WITH THE DISCOVERY of Neptune, the Solar System seemed to be complete once more. The wanderings of Uranus had been explained: the old observations of Flamsteed and Le Monnier had fallen into place, and all the irregularities which so puzzled Bouvard had been cleared up. Such was the general opinion for almost half a century. And then—very slowly, very slightly—Uranus started to wander again.

The differences between the predicted and actual positions were so small that they might have been due to observational errors. Yet it was possible that the cause was more fundamental, and Percival Lowell, best remembered for his theories about Mars but also a first-class mathematician, came to the conclusion that there must be another planet awaiting discovery. Accordingly, he began to work out where it ought to be, much as Adams and Le Verrier had done for Neptune.

Although the unknown planet was presumably well beyond Neptune's orbit, and would therefore affect Neptune more strongly than Uranus, Lowell preferred to base his calculations upon the wanderings of the inner of the two giants. This was because Neptune's movements were less well known. It had been discovered later, and had completed only one-third of a revolution round the Sun since Galle and d'Arrest had first identified it in 1846. Two pre-discovery positions recorded by the French astronomer Lalande in 1795 were available, but were of rather dubious accuracy.

Lowell was well equipped. He had built his observatory at Flagstaff, in Arizona, specially for planetary work, and he started hunting in 1905, though his final calculations were not published until 1914. 'Planet X', as he called the unknown body, was thought to lie just under 4,000 million miles from the Sun, moving in a rather eccentric orbit, and having a period of 282 years. As the effects upon Uranus were so small, Lowell thought that his X must be intermediate in size between a giant and a terrestrial planet, with a mass perhaps six times that of the Earth. Yet despite his careful searches, he was unsuccessful; and when he died, two years after the publication of his final memoir on the subject, Planet X was still unfound.

Meanwhile, the problem was being taken up by another American, W. M. Pickering. Unlike Lowell, Pickering decided to concentrate upon the movements of Neptune rather than that of Uranus, but he had another clue also, provided by those flimsy and erratic wanderers the comets.

Although sometimes of vast size, comets are not planet-like. They are made up of relatively small particles together with 'dust' and tenuous gas. Their masses are very small, negligible even compared with a small satellite such as Phœbe or Nereid, and their orbits can be violently disturbed by the gravitational pulls of the planets. A good case was provided by Lexell's Comet of 1770, which went so close to Jupiter that its orbit was completely changed.

Generally speaking, comets move in paths which are much more eccentric than those of any planet, and it is a striking fact that over fifty of them have their aphelia (maximum distance from the Sun) at about the mean distance of Jupiter. It seems that this is no mere coincidence, and astronomers still refer to Jupiter's 'comet family'. There may be similar families associated with the other giant planets, though the evidence is rather inconclusive.

Pickering pointed out that there were sixteen known comets with their aphelion distances at about 7,000 million miles from the Sun, and this made him even more certain that there must be a planet well beyond Neptune. His calculations led to a result similar to Lowell's, and in 1919 Milton Humason, at Mount Wilson, began to search.

Despite the probable faintness of Planet X, Humason's task was in some ways easier than that of Challis seventy years before. Challis had had to check each star visually, and the fact that he had no suitable chart was one of the reasons why he was not the first to identify Neptune (though, to be honest, he cannot be absolved from blame!). Humason, however, could make use of photography.

If an area of the sky is photographed twice, with an interval of one or two days between the exposures, the stars will stay in the same positions relative to each other, but a planet will be seen to have moved. All that need be done is to check the two photographs, and see whether any 'star' has shifted by the right amount. The process is immensely laborious, but it is at least straightforward.

To Pickering's disappointment, Humason was no more successful than Lowell had been, and after a time the search was given up. Nothing more happened for some years, but then, in 1929, astronomers at Flagstaff took up the problem again, armed with a 13-inch refractor and an ingenious instrument known as a blink-microscope for comparing exposed photographic plates. The programme was handed over to Clyde Tombaugh, now one of America's most senior and respected astronomers, but then a young research student. Tombaugh attacked his task with skill and energy, and in early 1930 he came across a suspicious object which proved to be the long-awaited planet.

Pluto, as it was named, was much fainter than anticipated, which is why it had evaded Lowell's own searches. Clyde Tombaugh told me a few months ago that the discovery was made sooner than he had dared to hope. 'When I embarked upon the search,' he said, 'I thought I would have to go all the way round the sky searching very systematically.' Yet Lowell's predictions had been very accurate indeed; and so, for that matter, had Pickering's. During the 1919 search, Milton Humason had photographed Pluto twice. One image fell upon a flaw in the plate, and the other was hopelessly masked by an inconvenient star, so that his failure to identify the new planet was sheer bad luck.

The trouble, all the way through, had been Pluto's unexpected faintness. Lowell had expected it to be of about the 12th magnitude, but in fact it was not much above magnitude 15. Otherwise it fitted the predictions quite well, even though it was closer to the Sun than Lowell had anticipated, and its revolution period was shorter.

But as information continued to come in, the whole situation became confused. It seemed that Pluto must be smaller than the Earth, and, presumably, less massive. If this were so, it could have virtually no measurable effects upon the movements of giants such as Uranus and Neptune—and yet it was by these alleged effects that Pluto had been tracked down. Could the discovery have been purely fortuitous?

Pickering believed that Pluto was not Lowell's Planet X at all. In 1928 he had made some new calculations, based largely on his comet studies, and had postulated the existence of another giant planet, larger than either Uranus or Neptune, which he called Planet P. He went so far as to work out a provisional orbit. Following Tombaugh's discovery in 1930, Pickering wrote:

> When I first recognized the importance of Planet P, from its comets, I mentally reserved for it the name Pluto, as the son of Saturn and the brother of Jupiter and Neptune; but unfortunately that small object now known as Pluto came round and perturbed Neptune some ten years before the leisurely P arrived and perturbed Uranus, and so received the name. Pluto should be renamed Loki, the god of thieves! A suitable name for P will indeed be difficult to find when that planet is discovered.

Now, forty years later, Planet P still eludes us—if it exists at all, which is far from certain—but Pluto is as mysterious as ever. Its orbit is remarkable. When at perihelion, Pluto is actually closer in to the Sun than Neptune can ever come, though for most of its revolution period of 248 years it is much more remote. The orbit is shown in Fig. 73. There is no danger of a collision with Neptune, since Pluto's path is inclined at the relatively sharp angle of 17 degrees.

What astronomers really wanted to have was an accurate value for the mass of Pluto, and, assuming normal density, the best method of finding out was to measure its diameter. Unfortunately even the Mount Wilson 100-inch reflector was unable to show a proper disk, and nothing much could be done until the Palomar 200-inch became available.

Kuiper made some preliminary measures in 1949, using the 82-inch reflector at the McDonald Observatory in Texas, and obtained a value of 6,400 miles, though with a large possible error. This would make the mass about 8/10 that of the Earth, so that the observed perturbations of Uranus and Neptune might just be accounted for, allowing for observational errors and a few coincidences.

In March 1950, twenty years after Tombaugh's original discovery, Humason and Kuiper, using the Palomar telescope,

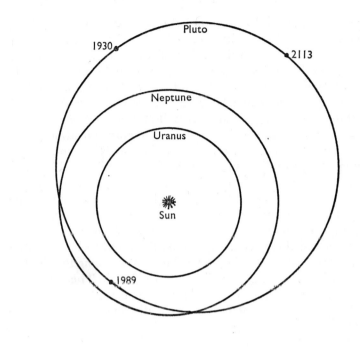

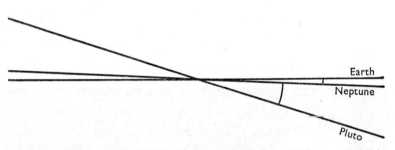

Fig. 73. *(Upper)* Orbits of Uranus, Neptune and Pluto. *(Below)* Inclinations of the orbits of Neptune and Pluto.

obtained the first measurements which might be expected to be reliable. The results were startling. Pluto proved to have a diameter of a mere 3,600 miles—less than that of Mars, and not much greater than that of Neptune's satellite Triton. It was thus smaller than any planet apart from Mercury.

Obviously, this threw the whole question wide open again. If Pluto were really smaller than Mars, and yet massive enough to cause the perturbations in the orbit of Uranus which had led to its discovery, its density would have to be about 12 times that of the Earth, or 60 times that of water. Here we run into immediate difficulties. Pluto would have to be made up entirely of very heavy materials, and with a tremendous surface gravity, so that a man who weighs 13 stone on earth would weigh over 60 stone on Pluto! A pull of this kind would indicate a high escape velocity, and consequently an atmosphere of hydrogen or helium which would be low-lying and dense, causing Pluto to reflect much of the feeble sunlight which falls upon it. (Most gases, of course, would liquefy, since the aphelion temperature must be in the region of −400 degrees Fahrenheit.) This does not agree with observation: it seems that Pluto's albedo, or reflecting power, is actually very low.

Neither did there seem any reason for Pluto to be so dense, and it was a distinct relief when later measurements put the diameter back to rough equality with the Earth. The perturbations could be explained—just. But then fresh measurements were made, and we were back to a diameter of only 3,700 miles (Fig. 74).

At present, this latter figure is the official one, so that if Pluto has a density equal to that of the Earth (5·5 times that of water) the mass is 0·1 that of Earth, and the escape velocity 3·2 miles per second. Clearly there are some odd aspects of the whole situation.

In 1936 A. C. D. Crommelin, at Greenwich, suggested the

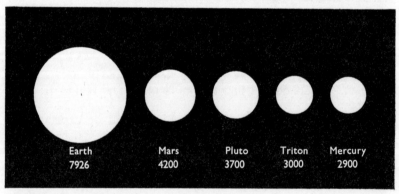

Earth	Mars	Pluto	Triton	Mercury
7926	4200	3700	3000	2900

Fig. 74. Relative sizes of the smaller planets of the Solar System, together with Triton. The diameter of Pluto is decidedly uncertain.

theory of 'specular reflection', which would mean that Pluto could be much larger (and more massive) than the diameter measurements indicate. If it had an atmosphere which cooled and froze, covering the surface with a smooth deposit, the planet would have a reflecting surface rather like that of a ball-bearing. Seen from sufficient distance, it would show a bright patch near its centre corresponding to the reflection of the Sun, surrounded by a darker area. We would see only the bright patch, and would be deceived into thinking that we were measuring the full diameter of the planet.

The whole idea hinges upon the albedo, which is generally thought to be only 14 per cent, less than that of Mars. It also seems improbable that a shiny surface could avoid being covered over with interplanetary dust. Crommelin's theory cannot be disproved, but few modern astronomers have much faith in it so far as Pluto is concerned. The discovery of a satellite of the planet would solve the mystery at once, but Pluto seems to be a solitary wanderer, and in any case a smaller satellite would be very hard to detect. Another suggestion is that Pluto may be simply the brightest of a trans-Neptunian group of asteroids, but there is no supporting evidence, and neither would it help us to solve the main puzzle of Pluto's mass.

The relatively eccentric, highly-inclined orbit is so strange that there have been doubts as to whether Pluto deserves to rank as a bona-fide planet. On one theory it is merely an ex-satellite of Neptune, which managed to break free and move off in an independent orbit.

In support of this idea there are two significant facts. First, Pluto is comparable in size with Triton; the diameter difference between the two may be only a few hundred miles. Secondly, Triton itself has retrograde motion, and is the only large satellite to behave in such a way.* All the other retrograde satellites—Phœbe in Saturn's system, and the four outer members of the Jovian family—are very small, and may be captured asteroids. Therefore Triton is unique, and there is no obvious reason why it should be so, unless it has been violently

* Since Uranus has an axial tilt of more than 90 degrees, and the five satellites move in its equatorial plane, their orbits may be classed as retrograde from a technical point of view, but are not generally regarded as being so. The whole case is entirely different, inasmuch as the satellites move round Uranus in the same direction as that of Uranus' axial rotation.

perturbed in the past. This could be connected with the escape of Pluto, and also, conceivably, with the comet-like orbit of Nereid. The actual process involved is not known, and, of course, the whole principle may be wrong; Pluto may never have been involved with the Neptunian system. But the idea is plausible in some ways.

It is further supported by the rotation period, which has been measured as being 6 days 9 hours. The discovery was made in 1955 by M. F. Walker and R. Hardie, using the 42-inch reflector at Flagstaff. They measured slight changes in brightness, which were regular, and about the same as those of Mars for one rotation: about one-tenth of a magnitude. There seems little doubt that they are due to Pluto's axial spin, and Walker and Hardie think that at the time of their observations Pluto was seen more nearly with its equator toward us than pole-on.

It is quite likely that most satellites have captured or synchronous rotation periods, as with our own Moon. Triton has a revolution period of 5·8 days, and probably a rotation period also of 5·8 days, so that if it were originally a quick spinner it has been slowed down by tidal interactions with Neptune. If the same happened to Pluto before the 'escape', we would expect a rotation period of the same order—which is precisely what we find. No other planet has a rotation period similar to this. Those of the inner planets Mercury and Venus are much longer; all the rest, much shorter.

Support is growing for Pickering's original idea that Pluto is not the planet responsible for the perturbations of Uranus and Neptune, but again we have little to guide us. Tombaugh's researches indicate that there is no trans-Plutonian planet brighter than magnitude 17; he could have picked up a planet the size of the Earth up to a distance two and a half times that of Pluto, and a Neptune-sized planet out to something like 20,000 million miles. However, any unknown planet is presumably much fainter than the seventeenth magnitude, and since there is no knowing exactly where it may be, even if it exists at all, a search would involve taking up the time of a giant telescope for a very long observational programme. This is hardly practicable, since the world's most powerful optical and photographic telescopes are already fully occupied.

One possibility occurs to me. If all goes well, a Grand Tour probe will by-pass Pluto during the 1980s, after having encountered Jupiter and Saturn. Once it has left the region of Pluto, it will continue in an orbit round the Sun, and signals from it will still be detectable. If an unknown planet happened to be in the right place at the right time, it could cause a marked perturbation in the orbit of the probe, and this could lead to the tracking-down of the planet itself.

I make the suggestion with great diffidence, because the odds against such a coincidence are very long indeed; but it is not absolutely out of the question, and it seems to hold out the only hope in the near future.

At present (1971) Pluto is still drawing inward toward the Sun. For some years to either side of its next perihelion passage, in 1989, it will be closer-in than Neptune, and will temporarily forfeit its title of 'the outermost planet'. Subsequently it will swing outward again; by 2113 it will have reached aphelion, four and a half thousand million miles from the Sun. For half a century around that date it will be so dim that only giant telescopes will show it, and, moreover, it will be so far south in the sky that it will never rise over Britain. Remember, Pluto's relatively tilted orbit means that unlike the other principal planets, it can leave the Zodiac. At the moment it is in Coma Berenices, which is not a Zodiacal constellation.

The magnitude is given officially as 14·9 to 15, but I suggest that this is wrong. Pluto is closer to us, and therefore brighter, than it used to be when Tombaugh first photographed it. I have seen it with my $12\frac{1}{2}$-inch reflector, and I estimate the magnitude as almost exactly 14; this has also been found by T. Moseley at Armagh, who has used the 10-inch refractor there, and whose results are probably better than mine. Of course, Pluto looks exactly like a star, and the only way to identify it is to compare it with the true stars close to it, tracking it down by means of its slow but perceptible motion from night to night. If Pluto would be obliging enough to occult a star, we might well obtain a better value for its diameter; as Taylor has done for Neptune; but the chances are slight, and no occultations are predicted for the foreseeable future.

Only the Palomar telescope will show a perceptible disk, and the apparent diameter is too small to be measured with precision. There is obviously no hope of seeing any surface details.

It has been said that the colour is slightly yellowish, but I am bound to be sceptical about reports of any colour in so dim an object.

Pluto must be a dismal world. From it, the Sun would appear only as large as Jupiter does to us, though it would still cast a brilliant light across the landscape. Neptune would be visible, and at times Uranus; but Plutonian astronomers—if they could exist!—would have a very poor view of the Solar System in general. Certainly the Earth could never be glimpsed.

The little we really know about Pluto tells us that it is the most desolate body in the Sun's family. It can never have known life; it marks the frontier of the Solar System.

Chapter Sixteen

BEYOND THE PLANETS

IT IS NOT ENTIRELY correct to say that the Solar System ends at the orbit of Pluto. Even if there are no additional planets, we must remember the comets, some of which swing out to tremendous distances, and take thousands or even millions of years to complete one circuit of the Sun.

Consider, for example, Bennett's Comet, which was quite bright for a few weeks during the early part of 1970. It arrived from the depths of space, passed round the Sun, and then receded once more. By now we have long since lost sight of it, and we will not see it again for many centuries. It will come back eventually, but not in our time; and there are some comets which may go out at least half-way from the Sun to the nearest star. Yet comets are now thought to be genuine members of the Solar System, though there is considerable argument as to the way in which they were formed.

If we disregard the great comets, the Solar System is very isolated. To recapitulate: represent the Earth–Sun distance by one inch, and the nearest star (Proxima Centauri) will be over four miles away. Most of the other stars are much more distant still, and no telescope yet built will show a star as anything but a point of light. (If you see a star as a large, shimmering disk, you may be sure that something is badly wrong with the focusing of your telescope!)

Remember that a normal star is very much larger than a planet; the Sun, which is officially ranked as a stellar dwarf, has ten times the diameter of Jupiter, which by planetary standards is decidedly large. In mass the discrepancy is even greater, since it would take 1,047 Jupiters to make up one body the mass of the Sun. It is important to note also, that there is a fundamental difference between a planet and a star, inasmuch as no planet is massive enough for nuclear reactions to be triggered off at its core.

Now look at the photograph in Plate XVI. Each tiny dot is a sun in its own right. If we looked at our Sun from an equivalent distance, we would clearly have no chance whatsoever of making out a body very close beside it, only one-tenth the diameter, and shining by reflection. It has been calculated that if Jupiter were moving round Alpha Centauri, the closest of the bright stars, it would have a magnitude of only +23. In fact, we cannot hope to detect planets of other suns by optical means.

Yet it is logical to assume that planet-families are common. The Sun is one of 100,000 million stars in our Galaxy; we can photograph many millions of similar galaxies, each with its own stellar quota. If, as is now believed, the planets in our Solar System were produced from a solar cloud, why should not the same process have been repeated elsewhere? At a conservative estimate, the number of planetary families in our Galaxy alone must run into tens of thousands, and probably into many millions.

Since no extra-solar planet can be observed, it might be thought that proof would be unobtainable. Luckily this is not so, and the problem has been attacked by an indirect method which has proved to be remarkably sensitive.

Because the stars are so remote, they show no individual or 'proper' motions marked enough to be noticed with the naked eye over periods of centuries. However, proper motions can be measured by modern techniques, and in some cases they are quite considerable. The holder of the apparent cosmical speed record is a faint Red Dwarf known as Barnard's Star. It is of magnitude 9·5, so that it is visible with a small telescope, and it lies in the constellation of Ophiuchus, the Serpent-bearer. Its distance is a mere 5·9 light-years, so that it is the closest known star apart from the three members of the Alpha Centauri system. It is moving across our line of sight at 55 miles per second, and it is approaching us at 67 miles per second (though there is no fear of an eventual collision; by the time it reaches our present position in the Galaxy, we shall be somewhere else). It appears dim because it is a celestial glow-worm with only 1/2300 of the Sun's luminosity.

Barnard's Star moves against the background of more remote stars by 10·31 seconds of arc per year, so that in 170 years it crosses a distance equal to the apparent diameter of the full moon. This may not seem very much, but it is fast by stellar standards, and has led to the nickname of the 'Runaway Star'. It is not really of exceptional velocity; its apparently rapid motion is due only to its closeness to us and the angle at which it is moving relative to the Sun and the Earth.

In 1937 Dr. Peter van de Kamp, Director of the Sproul Observatory in the United States, began to take a series of photographs of Barnard's Star. By the beginning of 1969 he had taken more than 3,000 plates on 766 nights, and the results were starting to look extremely interesting. Instead of moving in an apparently straight line against its background, Barnard's Star was 'wobbling'. Each 'wobble' was very slight —about 0·03 seconds of arc, so that measuring them on the photographic plates was like trying to detect a one-inch movement across a distance of a hundred miles. But it could be done, and at last van de Kamp was able to announce that Barnard's Star was being perturbed by an unseen companion with about $1\frac{1}{2}$ times the mass of Jupiter, moving round the star at an average distance of rather over 400 million miles in a period of 25 years.

A body only $1\frac{1}{2}$ times as massive as Jupiter cannot possibly be a star. Therefore, it must presumably be a planet. The only unexpected fact was that according to the calculations made by van de Kamp, the orbit would have to be very elliptical, more cometary than planetary in shape. This seemed improbable; and after making another series of calculations, van de Kamp modified his original announcement. Instead of one planet moving in an eccentric ellipse, it was more likely that there were two planets, provisionally designated B1 and B2.

On this new scheme, B1 moves round the star at a distance of 260,000,000 miles in a period of 12 years; B2 moves at 440,000,000 miles in a period of 26 years. Both orbits are circular, and in the same plane; both planets are comparable in mass with Jupiter. The diagram in Fig. 75 sets out the Barnard's Star system as it is now thought to be; the relative distances of the Earth and Jupiter from our Sun are given for comparison. The visual magnitudes of B1 and B2 are probably about +30, so that they would be well beyond the range of any existing telescope even if they were not so close to their parent star.

Searches of this kind are far from easy, and involved measuring distances on the photographic plates of the order of 1/25000

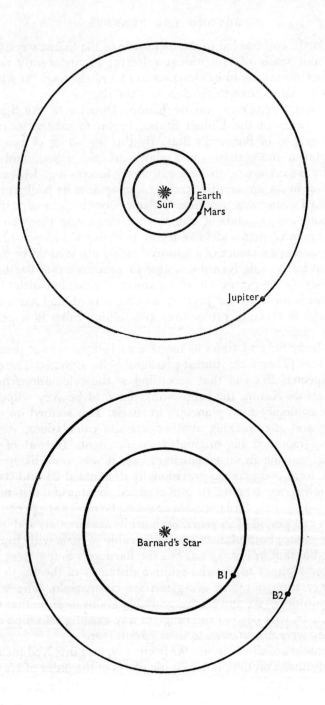

of an inch. A planet the mass of the Earth would be quite un-detectable, and the method is necessarily limited to the dis-covery of relatively massive planets associated with nearby, lightweight stars.

There had been several previous investigations of the same sort. In 1944 K. A. Strand had suggested that there might be a very massive planet moving round the fainter component of the binary star 61 Cygni, which is 11 light-years away, and is memorable as being the first star to have its distance measured (by F. W. Bessel, in 1838). According to Strand, the planet of 61 Cygni B was fifteen times as massive as Jupiter, which is perhaps disturbingly large. Another report concerned the star 70 Ophiuchi (no connection of Barnard's Star). But these cases are somewhat dubious, and the only system so far identified without question is that of Barnard's Star.

Van de Kamp is continuing his researches, and other close stars are under scrutiny. There are indications of an invisible body associated with Epsilon Eridani, a star 10·8 light-years away and not much smaller and cooler than our Sun; but the analysis is still incomplete, and it is too early to make any definite announcement.

This is as far as we can go at the moment; but the evidence, fragmentary though it may be, strongly favours the idea that planetary systems are common in the Galaxy, and no doubt in other galaxies as well. Though we cannot see extra-solar planets, we may be confident that they exist. We must remem-ber, too, that our Earth would be both invisible and undetect-able to beings who might live in (say) the system of Epsilon Eridani; even Jupiter would be hard to track down with in-struments of the kind we can contemplate. Looking at the Sun, our far-away astronomer could only guess that such a yellow, commonplace star might be the ruler of a system of nine planets, of which one is inhabited by beings who have already taken their first faltering steps into space.

Fig. 75. The system of Barnard's star, according to van de Kamp, compared with the Solar System, drawn to the same scale.

Chapter Seventeen

LIFE ON THE PLANETS

OF ALL THE PROBLEMS facing us, two are of special interest. First, what is the origin of the universe? And second, is there life anywhere beyond the Earth? If we could answer these questions, we would have taken a gigantic step forward.

The first problem is beyond the scope of this book, but there seems ample justification for saying something about the second, even though I cannot hope to provide a definite answer, and one's personal views are bound to enter into any discussion. As a preliminary, we must try to define what we mean by 'life'.

We do not know precisely what life really is; but we know a great deal about the structure of organic matter, and it has been found that all living material known to us depends upon the properties of the atom of carbon. Carbon atoms have a remarkable ability to build up complex atom-groups or molecules, both with other carbon atoms and with different elements; and it is these highly complicated molecules which are essential for life. The only other atom which has something of the same power is silicon, which, however, is not nearly so efficient; we have no evidence that silicon-based life can exist. All the other elements seem to be ruled out. This is not mere opinion; it is scientific fact, and mathematical analysis can prove it.

It follows that life, wherever it may be found, must be based upon carbon, unless other elements exist of which we have no knowledge. This latter idea flies in the face of all the evidence. There are 92 naturally-occurring elements found on Earth, and they form a complete sequence; there is no room for an extra one, and the additional elements that have been produced in our laboratories are very heavy and usually unstable, so that they do not enter seriously into the discussion. Moreover, spectroscopic analysis of remote stars and star-systems tells us that the elements there are just the same as the elements here. The light from the Pole Star or from an immensely remote galaxy is due to substances well known to us.

It is dangerous to be dogmatic. Science-fiction writers delight in producing B.E.M.s (Bug-Eyed Monsters), either made up of unknown elements or else based upon some familiar element such as iron or gold. Let us admit, at once, that it is impossible to disprove the existence of such alien creatures. On the other hand, it is equally true to say that if they do exist, then much of our modern science is wrong; and for this there is not the slightest evidence.

If we have a set of facts, all we can do is to take them and put the most reasonable interpretation on them—even if the set is incomplete. When we do this, we find that B.E.M.s do not fit into the picture. We are therefore entitled to discount them. If contrary evidence is produced in the future, then we must think again. This is not impossible, of course, but for the moment it is sensible to limit our discussion to 'life as we know it'. Once we go further, speculation becomes not only endless, but also rather pointless.

Life on Earth can assume many forms. There is not much superficial resemblance between an amoeba, a daffodil, a jelly-fish and a man; but the essential factor—dependence upon carbon—is there. Certainly all life of the kind we can under-stand requires the same sort of environment. There must be a reasonably equable temperature; a suitable atmosphere, and a supply of water. If any of these essentials is lacking, then there will be no life.

When we look at the planets in our Solar System, we find that most of them may be dismissed out of hand. Mercury has virtually no atmosphere, and a most uncomfortable tempera-ture-range. Jupiter and the other giants are too cold, as well as being devoid of solid surfaces (so far as we know); the idea of life-forms below the outer clouds seems very far-fetched. Pluto, even colder, is again minus any useful atmosphere. Of the asteroids and satellites, only Titan is definitely known to have an atmosphere, though Triton may also have; but a tenuous methane atmosphere is of no help at all, quite apart from the temperature drawback. As yet we have samples only from our own Moon, and these have shown a total lack of any sign of life either past or present, so that we may now be sure that the Moon has always been sterile.

Venus was regarded as a promising prospect until recently,

but it has now been found that the temperature there is extremely high, and the atmosphere is made up of carbon dioxide, producing a ground pressure of over 100 times that of our air at sea-level. So we come back, as always, to Mars.

In the last edition of this book, published in 1962, I wrote that 'it seems overwhelmingly probable that living organisms exist on Mars, and are responsible for the famous dark areas'. At the time this seemed logical; now, with regret, we must concede that the whole situation has changed. The Martian atmosphere is much more tenuous than we had thought, and the polar caps are unlikely to be made up of H_2O; it is not generally thought that the dark areas are of organic origin. Science is in some ways unromantic. Lowell's highly-civilized Martians were abandoned several decades ago, and it now seems that the Martian vegetation has followed suit.

Again we must be wary of dogmatism. There is always a chance that we are being over-pessimistic about both Mars and Venus, and we cannot discount primitive organisms there, perhaps underground in the case of Mars or in the higher, cooler atmosphere of Venus. Yet the odds are against anything of the sort, and I repeat that I am profoundly sceptical about the 'ammonia-based life' which, some astronomers have suggested, may exist beneath the clouds of the giant planets. I am open to conviction, but at the moment I simply do not believe in it.

The main point is that the Earth, and only the Earth, moves in the region which is known as the solar ecosphere—that is to say, the region in which the temperature is liable to be neither too hot nor too cold (Fig. 76). Venus is near the inner edge of the ecosphere, Mars at the outer boundary. Whether life has ever existed upon either planet remains to be seen, but I am ready to go on record as saying that I doubt it. Venus is a problem in every respect; its strange difference from the Earth must be due to its lesser distance from the Sun, but at any rate its whole evolution has been unlike ours. As for Mars—well, the low escape velocity presumably means that it lost the bulk of its atmosphere fairly quickly on the cosmical time-scale. It is not likely that life had the chance to develop there before conditions became hostile to it.

In the rest of the Solar System there seems, then, to be no

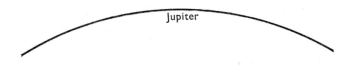

Jupiter

TOO COLD

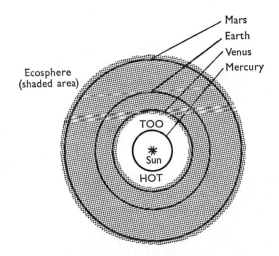

Ecosphere
(shaded area)

Mars
Earth
Venus
Mercury

TOO

✳
Sun

HOT

Fig. 76. The Ecosphere. Only the planets Venus, Earth and Mars lie within it, and only the Earth is almost central in it. A planet comparable with the Earth in size and mass would be too hot to support advanced life of our kind if it moved inside the ecosphere; if it lay beyond the ecosphere, it would be too cold.

life; certainly no intelligent life except (possibly) on the Earth. Yet this is not the same as claiming that we are unique. This would be a parochial view, and quite illogical.

There are many millions of stars which are very like the Sun. As we have noted, the incomplete evidence available to us indicates that many of these stars may have planets moving round them. And if we have a planet similar to the Earth, moving round a star similar to the Sun, it is logical to suppose that life will have appeared there. It may or may not be superficially like our own; I am prepared to believe that somewhere in the Galaxy there may be a highly intelligent astronomer with two heads and three arms. Yet he will not be a 'bug-eyed monster' in the true sense, because he, like ourselves, will be carbon-based, and he will need the same kind of environment.

Unfortunately, planets of other stars are so remote that they are out of contact with us. Sending a rocket probe out even to Alpha Centauri or Barnard's Star is absolutely out of the question at present, and will probably remain so, since nothing can travel faster than light (186,000 miles per second),* and any such journey at speeds we can contemplate would take an impossibly long time even if it were practicable in other ways. The only slim chance of establishing contact is by radio, since radio waves, which are electromagnetic vibrations, move at the same velocity as light.

In 1960 a team of scientists at Green Bank, West Virginia, headed by Frank Drake, carried out a series of experiments which would have seemed fantastic a couple of decades ago. We know that there are vast clouds of rarefied hydrogen spread between the stars of our Galaxy, and that these hydrogen clouds emit radiation at a wavelength of 21·1 centimetres. This 21-centimetre radiation can be collected by radio telescopes, such as the equipment at Green Bank, and has given us invaluable information about the structure of the Galaxy. If there are any other advanced beings anywhere, they will presumably know about the radio emission, and will be keeping a watch on this particular wave-length. The idea behind the Green Bank project was to try to detect signals at 21 centimetres which were sufficiently rhythmical to be classed as nonnatural, and could therefore be attributed to transmissions from some remote planet.

Drake concentrated upon the two nearest stars which are at all like the Sun. Both are in the southern sky; one is Tau Ceti, while the other is Epsilon Eridani, mentioned earlier as being a possible centre of a planetary system. It is quite on the cards that either or both is attended by a peopled planet, and so the

* Anyone who questions this must, I fear, start delving into the abstruse mathematics of relativity theory!

choice was logical. As well as 'listening out', the team transmitted signals which might be picked up and identified by
radio astronomers living in the Tau Ceti or Epsilon Eridani
systems.

The experiment was known officially as Project Ozma,
though I believe it was commonly nicknamed Project Little
Green Men! It was the longest of long shots; predictably, it
produced no positive results, and it was discontinued after a
few months. Yet it was anything but a waste of time, and it
may be revived and extended in the future. The chances of
contact are small, but not nil; and if we do not try, we cannot
hope for success.

Lastly, what of the more immediate future?

Manned bases will be set up on the Moon within the next
few years; of this there can be no reasonable doubt, always
provided that *homo insapiens* does not become involved in another global war which would certainly destroy civilization.
Mars is next on the list, and we may hope for bases there also
before the turn of the century. Unfortunately, life on the Moon
or Mars will always have to be under very artificial conditions;
the science fiction dream of providing Mars with a breathable
atmosphere is likely to remain a dream only, and it will take a
long time for the colony there to become self-supporting, if it
can ever do so. Neither can planetary colonization solve the
worst problem facing us today: over-population. Mars and
the Moon together could not support enough people to make
any difference to the Earth in this respect, and the solution
here must in the end be social rather than scientific, though I
admit that I have not the slightest idea of how to tackle it.

Further than this, we cannot see as yet. We must await the
results of work to be carried out within the next decade or two;
then, and then only, can we decide whether we must always
stay within our own local part of the Solar System. Meantime,
our task is to find out as much as we can about the worlds we
hope to explore. The planets are our companions in space; by
now they have become less remote and inaccessible, and they
offer us the greatest challenge ever made to the ingenuity of
mankind.

Appendix I

OBSERVING THE PLANETS

BEFORE THE SPACE AGE began, much of our knowledge of the surface features of the planets was based on amateur work. This was also true of the Moon. Things are, however, quite different today. There is still the same enjoyment to be gained from observing the planets and noting their special points of interest; surely nobody can tire of gazing at the rings of Saturn or the belts and satellites of Jupiter? But the amateur who wants to carry out useful research has to be more specialized and better-equipped than was the case even five years ago. As a general rule, I would say that he must have a telescope of at least 4 inches aperture (for a refractor) or 6 inches (for a reflector), and I would not personally be at all happy without something decidedly larger than this. Even the $12\frac{1}{2}$-inch and $8\frac{1}{2}$-inch reflectors at my private observatory at Selsey are adequate only for some branches of work; and few amateurs will be able to equip themselves with really complicated auxiliary instruments. Planetary spectroscopy, to give just one example, must be left in professional hands.

I do not propose to go into much detail here about observing methods, because I have done so elsewhere,* but a few hints will not be out of place. First, never try to use too high a magnification. A small, sharp image is always preferable to a larger, but blurred one. A magnification of about 50 per inch of aperture is about as much as can normally be used on good nights. On the other hand, I strongly disagree with the opinion that a small telescope can often show as much as a larger one. In my experience, the larger the telescope, the better the results (under good conditions, of course).

Secondly, do not hurry an observation, and always record it as soon as it has been completed. The temptation to 'leave it until tomorrow' should be stubbornly resisted, since mistakes are bound to creep in.

Thirdly, always write full notes. No observation is useful

* *The Amateur Astronomer* (7th edition: Lutterworth Press, 1971).

unless it carries the observer's name, the time (G.M.T.), telescope, magnification, and conditions of seeing. For the latter, the scale devised long ago by E. M. Antoniadi is suitable; it ranges from 1 (near-perfect conditions) down to 5 (very poor).

Next, never place any reliance upon an observation made under rushed or poor conditions. Such work is not only useless, but is actively misleading, as it may confuse subsequent analyses.

Last, and above all, never give way to wishful thinking. Record only what has been seen with certainty, and beware of jumping to conclusions. If, for instance, you go to the telescope confident that you will see the north polar cap of Mars, you will probably 'see' it whether it is there or not. Unconscious prejudice is difficult to avoid; the ability to do so is the supreme test of the observer's skill and experience.

With these comments, we may consider the planets one by one. I have given abbreviated notes; reference should be made either to the chapters in this book, or to the more detailed observational notes given in books written specifically for the purpose.

Mercury

No useful work with regard to the surface features of Mercury can be done without the use of a telescope quite beyond the amateur's range. One useful programme, however, is to estimate the phase to see whether there is any discrepancy similar to the Schröter effect with Venus. The best method is to make a drawing at the telescope, and then measure it; the results are much better than a simple estimate. The main trouble is that observing Mercury is rather pointless except in daylight or when the Sun has only just set (or has not quite risen). To locate Mercury under daylight conditions, one needs a powerful telescope equipped with accurate setting circles and clock drive. Never sweep aimlessly around searching for the planet; there is always the danger that the Sun will enter the field of view, with tragic results to the observer's eyesight. Binoculars are helpful in locating Mercury at dusk or dawn—but again, never sweep unless the Sun is actually below the horizon.

Venus

Despite the general paucity of surface detail, Venus is always a worthwhile object to observe. The main research programmes for the amateur are:

(i) Phase.—The Schröter effect, described in Chapter 6. As with Mercury, measure a drawing made at the telescope rather than be content with a straightforward estimate. We do not yet know nearly as much about the Schröter effect as we would like to do, and amateurs can play a useful rôle. Though the phase anomaly is most obvious near the time of dichotomy, it is generally present to some extent.

(ii) The shadings, which should be drawn as definitely as possible.

(iii) The brighter areas, including the persistent cusp-caps.

(iv) The shape of the terminator, which is by no means always a regular curve.

(v) Any sign of the Ashen Light. For this, it is essential to block out the bright crescent by some sort of occulting device fitted to the eyepiece, because the observations have to be made against a darkish background; Venus is then inconveniently bright, and is also rather low down, so that seeing conditions are unlikely to be good.

Dawn and dusk conditions are sometimes satisfactory, though many observers prefer to study Venus in broad daylight. Luckily the planet is relatively easy to find when well placed, and is visible in the average finder; but again, beware of sweeping around anywhere near the Sun. If you have an equatorial mounting, a quick method is to set the declination of Venus and swing *away* from the Sun by the difference in right ascension between the Sun and Venus.

Venus shows great variations in apparent diameter, which is at its maximum during the crescent stage; but in general it is best to make all drawings the same size. A diameter of 2 inches to the full circle is suitable.

Filters are an essential part of the equipment for the serious observer of Venus. Details are given in various books, and also in many papers in the *Journal of the British Astronomical Association* published between 1956 and the present time.

The Moon

Here, above all, the amateur's scope has been narrowed since the space probes were launched. It must be admitted that mapping the surface features with a small or even a large telescope is now something to be undertaken for pleasure only; but the would-be observer must start by learning his way around, and drawing the various craters and other features is the best way to do it. Obtain an outline lunar map, and work through it, bearing in mind that a crater which is prominent when shadow-filled at the terminator may be unidentifiable under high illumination, and that for most purposes full moon is the worst possible time to start observing.

Certain relatively inconspicuous features, such as domes, can be usefully studied; the Orbiter photographs cannot cover the Moon under all conditions of illumination, so that in some of the pictures the very low and gentle features do not show up. Then, of course, there are the T.L.P.s, or Transient Lunar Phenomena, which take the form of short-lived reddish glows, indicating some sort of mild surface activity. Amateurs have played a decisive part in this research. A telescope of at least 6 inches aperture is needed, preferably with a system of rotating filters (Moon-blink device). The main hazard is, as always, unconscious prejudice. If you think you have noted an unusual coloration, check all adjacent areas to make sure that you have not been deceived by some effect due to the Earth's atmosphere or instrumental trouble; and check the position and time very precisely. No T.L.P. report is really valuable unless confirmed independently by another observer at a separate site. And in this work, a few faulty reports can wreak havoc with all subsequent attempts at analysis.

Mars

Mars is a difficult object for a small aperture. It comes to opposition only in alternate years, and in general useful observations can be made only for a few months to either side of the opposition date. One really needs a telescope of over 8 inches aperture for valuable work, though of course the main features can be seen with a much smaller aperture.

Owing to the smallness of the disk, Mars is the one planet upon which a really high power should be used whenever at all possible. It is best to select a definite scale for the drawings (preferably 2 inches to the planet's diameter), and the phase, which is often considerable, should never be neglected. This can be worked out beforehand from yearly tables, and the disk outline prepared accordingly.

Before starting to draw Mars, it is wise to spend some time in surveying the planet, until the eye has become thoroughly adapted. When the drawing is begun, the polar cap and main dark areas should be sketched in as quickly as is compatible with accuracy, because Mars is rotating on its axis and there is a slow but perceptible drift of the markings across the disk. Minor details can then be added at leisure. When the drawing is complete, and you are satisfied that nothing has been missed, written notes should be added about colours, intensities, and any features of special interest, notably clouds.

Do not expect to see too much. At first you may be able to make out nothing apart from a few dark areas and the white polar cap; but as you gain experience, you will see more and more.

The Minor Planets

Here, the main interest for the amateur lies in identifying the minor planets and following their movements from night to night. One (Vesta) can sometimes be seen with the naked eye, and several more are within the range of binoculars and small telescopes. Their positions can be found from the *Handbook* of the British Astronomical Association, together with a star atlas (Norton's is the best). With a little practice they may be found quite easily, and estimates of their magnitudes are by no means devoid of value.

Jupiter

It is no exaggeration to say that most of our long-term knowledge of Jupiter's surface features is due to amateur astronomers. The disk abounds in detail, and a modest telescope of aperture 6 inches or so is enough for useful work to be done.

Disk drawings should be made as quickly as possible, as the rate of spin is rapid and the drift of the markings is very obvious. A drawing should be completed within ten minutes at most with regard to the major details; minor ones can be filled in more slowly, without altering the general framework. The colours seen should then be noted, and also the relative intensities of the various belts and zones.

A very important part of the Jupiter programme is the taking of transits. A feature transits when it passes across the planet's central meridian, i.e. the line passing through the two poles and the centre of the disk. The polar flattening enables the central meridian to be found easily—and, incidentally, this flattening should never be neglected when a drawing is to be made. Personally, I use prepared printed blanks.

Transit times should be estimated to the nearest minute, and accurate timing is essential. It might seem a difficult task to make the estimates with sufficient precision, but with a little practice it becomes strangely easy, and it is possible to take many transits in the course of an hour or so as the markings drift steadily across the yellowish Jovian disk.

These transits are important because they allow the longitude of the feature to be calculated, and hence rotation periods to be derived. As has been shown, these periods are of interest to both optical and radio astronomers. It is easy to work out the longitudes from the tables given in the *Handbook* of the British Astronomical Association, which give the longitude of the central meridian for every hour; the only arithmetic involved is simple addition and subtraction.

As the equatorial zone (System I) rotates more quickly than the rest of the planet (System II), two sets of tables are necessary, and there must be no confusion as to which set is to be used—otherwise the results will be most peculiar! (Remember that System I is bounded by the northern edge of the South Equatorial Belt and the southern edge of the North Equatorial Belt.) It is also necessary to indicate in which part of the belt the feature lies; for instance, NEB(s) indicates the southern part of the North Equatorial Belt. The usual abbreviations are: P = polar, T = temperate, Tr = tropical, Eq = equatorial, B = belt, Z = zone, D = disturbance, RS = Red Spot, pr = preceding, f = following.

It may also be useful to give an extract from my own observing notebook for one particular night:

'1970 April 13/14. 12½ in. reflector, × 360. Seeing 3 till

00.02; thereafter 3 to 4. NEB broad and dark, and the STB also prominent, but the SEB was faint and double. The NTB was visible, and a darker section followed the transit at 00.28 (April 14). The Red Spot was much in evidence, and was highly coloured, but no sign of the Hollow. A bright narrow zone lay between the STB and the SSTB.

Transits:		Longitude:	
		I	II
23.21	(Apr. 13). Pr. end of the Red Spot	. . .	017·9
23.25	F. end of projection from S. edge of NEB	023·9	. . .
23.34	Centre of Red Spot	. . .	025·7
23.37	White spot in N. of EqZ	031·2	. . .
23.41	Condensation in SSTB	. . .	029·9
23.46	F. end of Red Spot	. . .	032·9
00.02	(Apr. 14) Projection from S. edge of STB	. . .	042·5
00.28	Pr. of dark section of NTB	. . .	058·2
02·03	Projection from S. edge of NEB	120·1	. . .
02.18	White spot in N. of EqZ	129·2	. . .

(Clouds, 23.48 to 00.00, 00.10 to 00.15, and 00.30 to 01.50.) Occultation of Io.—1st contact, 23.54. Vanished, 23.56, Emersion, 02.15 (definition poor at this time).
Clouds stopped observing for the night at 02.20.'

I do not claim that this was a good series of observations; I give it as a typical example of work on a rather poor night. I made a drawing at 23.37–23.47, and a strip drawing of the EqZ at 02.00–02.10.

So far as the Galilean satellites are concerned, the most useful work is to make careful timings of the transits (immersion and emersion at the planet's limbs), shadow transits, occultations and eclipses. These observations help in improving our knowledge of the movements of the satellites.

Saturn

Saturn is in some ways a convenient planet. It bears high powers well, even better than Jupiter; and although there is not generally much surface detail, there is always the chance of making a startling discovery—as W. T. Hay did on August 3, 1933, when he detected the famous white spot.

The paucity of well-defined detail means that surface transits arc difficult to take, but whenever possible they should be observed in the same way as for Jupiter. For this work, a telescope of at least 10 inches aperture is desirable. It is also valuable to estimate the colours and intensities of the various zones of the planet, as marked, and so far unexplained changes occur from time to time.

Cassini's Division is easy when the rings are fairly well open, as in the early 1970s; but Encke's is elusive, and should be looked for near the ansæ. By all means look for the unconfirmed dusky ring outside Ring A, though I am very dubious as to whether anyone will see it. The reported new ring inside the Crêpe Ring is also worth a search, though the only real method is to check the intensity of the blackness in the 'gap' against that of the outer sky. Estimate the colours and intensities of the rings, comparing them with each other and also with the disk, using a scale of from 0 (white) to 10 (black shadow). When a drawing is made, take great care to put in the shadows very accurately: globe upon rings, rings upon globe. The shadow on the rings is important, because any irregularities may give a clue as to the structure of the rings themselves. Saturn is really at its most interesting, though least beautiful, when the rings are edgewise-on, but this will not occur again until 1980.

Occasionally Saturn passes in front of a star, and this is an important event, since it enables the transparencies of the various rings to be determined.

Any 3-inch refractor will show Titan well, and with my 12½-inch reflector I have managed to see all the satellites except for Phœbe. The main work with regard to the satellites is in magnitude estimation. Convenient stars may be used as comparisons (be sure to identify them correctly!) but when there are no stars to hand, as often happens, the only real solution is to adopt a constant magnitude of 8·3 for Titan, and work from that. Errors are bound to be introduced, but the method is better than nothing. Iapetus, of course, is of special interest.

Uranus

There should be no difficulty in locating Uranus, since even binoculars will show that it presents a small disk. The most useful amateur work is to estimate its magnitude, which, as we have noted, seems to show curious fluctuations. A low power is advisable, since with higher magnifications Uranus becomes so un-stellar in aspect that it cannot be compared with a star so far as magnitude is concerned. The path of the planet for each year is given in the B.A.A. *Handbook* and also in the annual *Yearbook of Astronomy*.

Neptune

Neptune may be found with binoculars, but even large telescopes are unable to show much in the way of surface detail. Its magnitude may be estimated in the same way as for Uranus, though Neptune appears to be much less subject to fluctuations.

Pluto

Pluto can now be glimpsed with a 12-inch telescope, but is difficult to identify, as it looks just like a very faint star. Over a few nights, however, its slow but perceptible motion will betray it.

I have said nothing here about amateur photography of the planets, which is admittedly rather difficult even though useful results may be obtained with Venus and Jupiter. The Moon, of course, is an ideal photographic subject, always provided that the observer is equipped with a proper astronomical camera together with an equatorial, clock-driven telescope.

Appendix II

ASTRONOMICAL SOCIETIES

ANY SERIOUS AMATEUR will be wise to join an astronomical society. Lone observing is enjoyable, but working together with others of like interest is much more so; and for any useful research, co-operation is essential.

The leading amateur society in Britain is the British Astronomical Association (Burlington House, London W.1). There are special sections devoted to the Moon; Mercury and Venus; Mars; Jupiter, and Saturn, each controlled by an experienced Director. No qualifications other than interest and enthusiasm are needed for membership. Meetings are held monthly at Burlington House (October–June), and there are various publications, including a bi-monthly *Journal* and an annual *Handbook*. The B.A.A. has an observational record second to none.

There are many local societies of high standard; for instance in Edinburgh, Birmingham, Norwich, Liverpool, Manchester, Portsmouth and elsewhere. A full list is given in the annual *Yearbook of Astronomy*.

Appendix III

LITERATURE

THE FOLLOWING LIST is very brief. I have given only books which are comparatively specialized, and are still in print.

(a) Books
ALEXANDER, A. F. O'D. *The Planet Saturn.* London, 1962.
—— *The Planet Uranus.* London, 1967.
FIELDER, G. *Lunar Geology.* London, 1965.
FIRSOFF, V. A. *The Interior Planets.* Edinburgh, 1968.
—— *The World of Mars.* Edinburgh, 1969.
MOORE, P., and CATTERMOLE, P. J. *The Craters of the Moon.* London, 1967.
PEEK, B. M. *The Planet Jupiter.* London, 1961.

(b) General Information
various.   *Yearbook of Astronomy.* London, annually.
various.   *Practical Amateur Astronomy,* London, 1970.

(c) Periodicals
Sky and Telescope: Harvard (U.S.A.), monthly.
Astronomy and Space: David & Charles (Newton Abbot), quarterly.

(d) Atlases, etc.
HATFIELD, H. R. *The Amateur Astronomer's Photographic Lunar Atlas.* London, 1968.
MOORE, P. *24-inch Map of the Moon.* Tunbridge Wells, 1970.

Appendix IV

(i) PLANETARY DATA

Planet	Mean distance from Sun, miles	Sidereal Period	Orbital Eccentricity	Orbital Inclination	Mean Orb. Velocity mi/sec.
MERCURY	36,000,000	88·0 days	0·206	7° 00′	29·8
VENUS	67,200,000	224·7 ,,	0·007	3 24	21·8
EARTH	92,957,000	365·3 ,,	0·017	—	18·5
MARS	141,500,000	687·0 ,,	0·093	1 51	15·0
JUPITER	483,300,000	11·86years	0·048	1 18	8·1
SATURN	886,100,000	29·46 ,,	0·056	2 29	6·0
URANUS	1,783,000,000	84·01 ,,	0·047	0 46	4·2
NEPTUNE	2,793,000,000	164·79 ,,	0·009	1 46	3·4
PLUTO	3,666,000,000	248·43 ,,	0·248	17 10	2·9

Planet	Axial Rotation (Equatorial)	Axial Inclination degrees	Mean Synodic Period days	Albedo	Maximum Magnitude at mean opposition
MERCURY	58·7 days	?	115·9	0·06	0·0
VENUS	243 ,,	?	584·0	0·76	− 4·4
EARTH	23h 56m 04s	23° 27′	—	0·36	—
MARS	24h 37m 23s	23 59	779·9	0·16	− 2·0
JUPITER	9h 50m 30s	3 04	398·9	0·73	− 2·6
SATURN	10h 14m	26 44	378·1	0·76	+ 0·7 .
URANUS	10h 49m	97 53	369·7	0·93	+ 5·7
NEPTUNE	About 14h	28 48	367·5	0·84	+ 7·9
PLUTO	6d 9h	?	366·7	0·1?	+14·0

Planet	Equatorial Diameter miles	Escape Velocity mi/sec.	Density: water = 1	Volume: Earth = 1	Mass: Earth = 1	Surface Gravity: Earth = 1
MERCURY	2,900	2·6	5·5	0·06	0·06	0·38
VENUS	7,700	6·4	5·3	0·86	0·82	0·90
EARTH	7,927	7·0	5·5	1	1	1
MARS	4,200	3·2	3·9	0·15	0·11	0·38
JUPITER	88,700	37·1	1·3	1,319	318	2·64
SATURN	75,100	22·0	0·7	744	95	1·16
URANUS	29,300	13·9	1·7	47	15	1·11
NEPTUNE	31,500	15·5	1·8	54	17	1·21
PLUTO	3,700?	?	?	+0·1?	?	?

Planet	Oblateness	Max surface temperature, degrees F.	Number of Satellites	Apparent Diameter: max. min. (seconds of arc)	
MERCURY	0·0	+770	0	12·9	4·5
VENUS	0·0	+887	0	66·0	9·6
EARTH	0·003	+140	1	—	—
MARS	0·005	+80	2	25·7	3·5
JUPITER	0·062	−200	12	50·1	30·4
SATURN	0·096	−240	10	20·9	15·0
URANUS	0·06	−310	5	3·7	3·1
NEPTUNE	0·02?	−360	2	2·2	2·0
PLUTO	?	?	0	?	?

(ii) SATELLITE DATA

Satellite	Mean distance from centre of primary, thousands of miles	Sidereal Period d. h. m.			Orbital Eccentricity	Orbital Inclination (To equator of primary) degrees
EARTH						
Moon	239	27	7	43	0·055	23·4
MARS						
Phobos	5·8		7	39	0·021	1·1
Deimos	14·6	1	6	18	0·003	1·8
JUPITER						
Amalthea (V)	113		11	57	0·003	0·4
Io (I)	262	1	18	28	0·000	0·0
Europa (II)	417	3	13	14	0·0001	0·5
Ganymede (III)	666	7	3	43	0·001	0·2
Callisto (IV)	1,170	16	16	32	0·007	0·2
Hestia (VI)	7,120	250	26		0·158	28
Hera (VII)	7,300	259	16		0·207	28
Demeter (X)	7,300	260	12		0·107	29
Adrastea (XII)	13,000	*631			0·169	147
Pan (XI)	14,000	*692			0·207	163
Poseidon (VIII)	14,600	*744			0·410	148
Hades (IX)	14,700	*758			0·275	157
SATURN						
Janus	98		17	58	0·00?	0·0?
Mimas	113		22	37	0·020	1·5
Enceladus	149	1	8	53	0·005	0·0
Tethys	183	1	21	18	0·000	1·1
Dione	235	2	17	41	0·002	0·0
Rhea	328	4	12	25	0·001	0·3
Titan	760	15	22	41	0·029	0·3
Hyperion	920	21	6	38	0·104	0·6
Iapetus	2,200	79	7	56	0·028	14·7
Phœbe	8,050	*550	10	50	0·163	150

SATELLITE DATA

URANUS

Miranda	76	1	9	50	0·000	0·0
Ariel	119	2	12	29	0·003	0·0
Umbriel	166	4	3	28	0·004	0·0
Titania	272	8	16	56	0·002	0·0
Oberon	364	13	11	7	0·001	0·0

NEPTUNE

Triton	220	*5	21	3	0·000	160
Nereid	3,500	360			0·749	27·7

Satellite	Diameter, miles	Density, water = 1	Maximum Magnitude	Reciprocal Mass, Primary = 1	Discoverer

EARTH

Moon	2,160	3·3	−12·7	81·3	—

MARS

Phobos	14·5	?	11	?	Hall, 1877
Deimos	7	?	12	?	Hall, 1877

JUPITER

Amalthea	150	?	13	?	Barnard, 1892
Io	2,310	4·1	5·5	26,200	Galileo, 1609
Europa	1,950	3·7	5·7	40,300	Galileo, 1609
Ganymede	3,200	2·4	5·1	12,200	Galileo, 1609
Callisto	3,000?	2·0	6·3	19,600	Galileo, 1609
Hestia	100	?	13·7	?	Perrine, 1904
Hera	35	?	17	?	Perrine, 1905
Demeter	15	?	18·8	?	Nicolson, 1938
Adrastea	14	?	18·9	?	Nicolson, 1951
Pan	19	?	18·4	?	Nicolson, 1938
Poseidon	35	?	18	?	Melotte, 1908
Hades	17	?	18·4	?	Nicolson, 1914

SATURN

Janus	150?	?	14	?	Dollfus, 1966
Mimas	300?	1	12	15,000,000	Herschel, 1789
Enceladus	400?	1	11	7,000,000	Herschel, 1789
Tethys	700?	1·1	10·5	910,000	G. D. Cassini, 1684
Dione	900?	3·2	10·4	490,000	G. D. Cassini, 1684
Rhea	1,100?	2	9·3	250,000	G. D. Cassini, 1672
Titan	3,300?	2·3	8·3	4,150	Huygens, 1655
Hyperion	200?	3	13	5,000,000	Bond, 1848
Iapetus	1,500?	?	9	?	G. D. Cassini, 1671
Phœbe	150?	?	14	?	Pickering, 1898

URANUS

Miranda	200	5	17	1,000,000	Kuiper, 1948
Ariel	1,500	5	14	67,000	Lassell, 1851
Umbriel	800	4	14·7	170,000	Lassell, 1851
Titania	1,500	6	14	20,000	Herschel, 1787
Oberon	1,500	5	14	34,000	Herschel, 1787

NEPTUNE

| Triton | 3,000? | 5 | 13 | 750 | Lassell, 1846 |
| Nereid | 200 | ? | 19 | ? | Kuiper, 1949 |

In this table, an asterisk indicates retrograde motion; whether the five satellites of Uranus should be included in this category is open to doubt, since their direction is the same as that of the axial rotation of Uranus itself.

Many of the figures given in the table are highly uncertain, particularly with regard to the diameters, densities and even magnitudes of the smaller satellites of the outer planets. Different authorities give different values. All we can really say is that generally speaking, the figures given here are of the right order.

(iii) MINOR PLANET DATA

The First Ten Minor Planets

Asteroid	Year of Discovery	Sidereal Period, years	Mean Distance from Sun, thousands of miles	Orbital Inclination	Diameter, miles	Max. Mag. (mean opp.)
1 Ceres	1801	4·60	257·0	10° 36′	427	7·4
2 Pallas	1802	4·61	257·4	34 48	280	8·7
3 Juno	1804	4·36	247·8	13 00	150	8·0
4 Vesta	1807	3·63	219·3	7 08	370	6·0
5 Astræa	1845	4·14	239·3	5 20	111	9·9
6 Hebe	1847	3·78	225·2	14 45	106	8·5
7 Iris	1847	3·68	221·5	5 31	93	8·7
8 Flora	1847	3·27	204·4	5 54	77	9·0
9 Metis	1848	3·69	221·7	5 36	135	8·3
10 Hygeia	1849	5·59	292·6	3 49	220	9·5

Some Close Approach Minor Planets

Asteroid	Year of Discovery	Sidereal Period, years	Orbital Inclination, degrees	Orbital Eccentricity	Distance from Sun in Astronomical Units max.	Distance from Sun in Astronomical Units min.	Diameter, miles
433 Eros	1898	1·76	10·8	0·22	1·81	1·11	17
719 Albert	1911	4·16	10·8	0·54	3·98	1·19	2
887 Alinda	1918	4·00	9·0	0·54	3·88	1·16	2
1221 Amor	1932	2·67	11·9	0·44	2·75	1·09	5
— Apollo	1932	1·81	6·4	0·57	2·33	0·64	1½
— Adonis	1936	2·76	1·5	0·78	3·50	0·43	1
— Hermes	1937	1·47	4·7	0·47	1·90	0·68	1
1566 Icarus	1949	1·12	23·0	0·83	1·97	0·19	1
1620 Geographos	1954	1·39	13·3	0·34	1·66	0·83	1
1580 Betulia	1952	3·25	52·0	0·49	3·28	1·11	5
1627 Ivar	1957	2·53	8·4	0·40	2·59	1·14	3

The Names of the First Hundred Minor Planets

1	Ceres	35	Leucothea	69	Leto
2	Pallas	36	Atalanta	70	Panopea
3	Juno	37	Fides	71	Feronia
4	Vesta	38	Leda	72	Niobe
5	Astræa	39	Lætitia	73	Clytie
6	Hebe	40	Harmonia	74	Galatea
7	Iris	41	Daphne	75	Eurydice
8	Flora	42	Isis	76	Freia
9	Metis	43	Ariadne	77	Frigga
10	Hygeia	44	Nysa	78	Diana
11	Parthenope	45	Eugenia	79	Eurynome
12	Victoria	46	Hestia	80	Sappho
13	Egeria	47	Melete	81	Terpsichore
14	Irene	48	Aglaia	82	Alcmene
15	Eunomia	49	Doris	83	Beatrix
16	Psyche	50	Pales	84	Clio
17	Thetis	51	Virginia	85	Io
18	Melpomene	52	Nemausa	86	Semele
19	Fortuna	53	Europa	87	Sylvia
20	Massilia	54	Calypso	88	Thisbe
21	Lutetia	55	Alexandra	89	Julia
22	Calliope	56	Pandora	90	Antiope
23	Thalia	57	Mnemosyne	91	Ægina
24	Themis	58	Concordia	92	Undina
25	Phocæa	59	Elpis	93	Minerva
26	Proserpine	60	Echo	94	Aurora
27	Euterpe	61	Danaë	95	Arethusa
28	Bellona	62	Erato	96	Ægle
29	Amphitrite	63	Ausonia	97	Clotho
30	Urania	64	Angelina	98	Ianthe
31	Euphrosyne	65	Cybele	99	Dike
32	Pomona	66	Maia	100	Hekate
33	Polyhymnia	67	Asia		
34	Circe	68	Hesperia		

INDEX

Sun, 12
distance of, 12, 122
energy of, 12, 25
future of, 25–6
origin of, 19, 23
rotation of, 20
status of, 12, 191
Supernovæ, 23
Surveyor-3, 94, 98
Swift, L., 48
Syrtis Major, 105, 110

T.L.P.s, see Moon
Tau Ceti, 200–1
Taylor, G., 177–8, 189
Telescopes, invention of, 16
suitable for planetary observation, 203
Telstar, 74
Tethys, 156, 159
Thebit (lunar crater), 94
Themis, 161
Thetis, 119
Thule, 118, 121, 123
Tidal friction, 54
theories, 20–2
Tides, the, 73–4
Titan, 83–4, 157, 161, 197, 210
Titania, 171–2
Titius, 115
Tombaugh, C., 75, 183
Transits of inferior planets, see Mercury and Venus
Triton, 83–4, 179–80, 187, 197
Troposphere, the, 78
Trubitsyn, V., 167, 179
Tsiolkovskii, K. E., 39–40
Trafton, L., 132–3
Troilus, 125
Trojan asteroids, 118, 125–6
Twinkling of stars and planets, 27
Tycho Brahe, 30
Tycho (lunar crater), 86, 93, 96

Umbriel, 171–2
Uranium, decay of, 18
Uranus, 11, 14–15, 19–20, 116, 164–73
axial inclination of, 168–9, 187
calendar of, 168
composition of, 167
conjunction with Jupiter, 170
data, 166
discovery of, 164–5
magnitude of, 166, 169–70, 210–1
naming of, 165
Neptune, perturbations by, 174–5
observing hints, 210–1
occultations by, 170
probes to, 43, 172
rotation of, 168, 187
satellites of, 171–2, 187

surface features of, 169
temperature, 166
view from, 173
Ussher, Archbishop, 17

V.2 rockets, 39
Valz, B., 119
Van Allen, J., 79
Van Allen Zones, 79
Van de Kamp, P., 193, 195
Van den Bos, 122
Venera probes, 66–7
Venus, 11, 14–16, 27–8, 59–71, 205
atmosphere, 61, 63, 66–7
brilliance, 59
calendar, 69–70
clouds, 62, 63
dimensions, 59
life on?, 63, 197–8
linear features reported on, 106
magnetic field, 63, 65
maps of, 62
observing hints, 205
occultations by, 62
orbit, 31
phases of, 34, 59–61
photography of, 62
radar maps of, 68–9
rockets to, 38, 41–2, 44, 63–4, 66–7
rotation of, 63–4, 67, 69–70
Schröter effect, 60–1
surface conditions, 69–71
surface markings, 62, 106
temperature, 63, 67–8
transits of, 34, 59, 61–2
water on?, 63
Von Bülow, 95
Von Rothschild, 120
Von Weizsäcker, C., 24
Von Zach, F. X., 116–17, 165
Vulcan, 47–8

Walker, M. F., 188
Watson, 48
Wells, R. A., 110
Whipple, F. L., 160
Wildt, R., 132–4
models by, for giant planets, 132–4, 149, 167, 178
Wilson, 171
Witt, 121
Wolf, Max, 118–19, 123
Wright, Orville and Wilbur, 39
Wright, T., 19

X (Planet), 181, 182, 184

Zharkov, V., 167, 179
Zodiac, 35–6
Zodiacal Light, 78